Fast Pyrolysis of Technical Lignins in a Circulating Fluidized Bed Reactor

Vom Promotionsausschuss der
Technischen Universität Hamburg
zur Erlangung des akademischen Grades
Doktor-Ingenieur (Dr.-Ing.)

genehmigte Dissertation

von

Miika Franck

aus

Hamburg

2019

Bibliografische Information der Deutschen Nationalbibliothek
Die Deutsche Nationalbibliothek verzeichnet diese Publikation in der
Deutschen Nationalbibliografie; detaillierte bibliographische Daten
sind im Internet über http://dnb.d-nb.de abrufbar.
    1. Aufl. - Göttingen: Cuvillier, 2019
    Zugl.: (TU) Hamburg, Univ., Diss., 2019

Date of oral examination: Mai 2nd, 2018

Committee
First examiner: Prof. Dr.-Ing. habil. Dr. h.c. Stefan Heinrich
Second examiner: Prof. Dr.-Ing. habil. Irina Smirnova
Committee chair: Prof. Dr.-Ing. habil. Martin Kaltschmitt

© CUVILLIER VERLAG, Göttingen 2019
   Nonnenstieg 8, 37075 Göttingen
   Telefon: 0551-54724-0
   Telefax: 0551-54724-21
   www.cuvillier.de

   ISBN 978-3-7369-9965-7
   eISBN 978-3-7369-8965-8

to my family

I owe it all to you

Acknowledgements

I would like to express my thanks to Prof. Joachim Werther, who initiated the pyrolysis research at the Institute of Solids Process Engineering and Particle Technology. Further I would like to thank committee chair Prof. Martin Kaltschmitt and committee members Prof. Stefan Heinrich and Prof. Irina Smirnova. In particular I highly appreciate Prof. Smirnova for traveling to Juelich to fight for project extension while in her advanced pregnancy.

I would like to express my sincere thanks to my colleagues of the Biorefinery2021 research cluster, which was funded by the German Ministry of Education and Research (BMBF). Especially to mention are Patrick Eidam, Ingrid Fortmann, Silke Radtke and Michael Windt of the Thuenen Institute of Wood Research, who supported me with measurement of the oil composition, sharing their knowledge in lignin chemistry and pyrolysis. Same holds for their team leader Dr. Dietrich Meier, to whom I express my deepest appreciation for reviewing this work and providing me with valuable feedback. For producing the hydrolysis lignin and valuable help with the adsorption measurements I am indebted to Dr. Carsten Zetzl, Dr. Lilia Zenker, Dr. Christian Kirsch, Lisa Marie Schmidt and Wienke Reynolds of the Institute of Thermal and Separation Processes, Hamburg University of Technology.

I am also very grateful to Dr. Ernst-Ulrich Hartge for many fruitful discussions concerning this work. I would like to express my gratitude towards Frank Rimoschat, Heiko Rohde, and Bernhard Schult for setting up the plant including measurement instrumentation and for support during the experiments. In addition a thank-you goes to my fellow doctoral students at the Institute for their high-value feedback, sharing many great moments, and overcoming numerous obstacles together. Especially to name are my office mates: Dr. Marvin Kramp, Dr. Andreas Thon, Dr. Johannes Neuwirth, Britta Buck and Monika Goslinska.

I would also like to express my sincere appreciation to the students involved in my research project: Philipp Clauss, Martina Holz, Karsten Gescher, Simon Sauerschell, Tom Wytrwat, Tobias Becke, Jan-Malte Bettien, Anna-Lena Bologna, Sergey Chernikov, Hanna Evers, Jonathan Faass, Lars Groos, Martin Hug, Jan-Philipp Kreienborg, Arthur Krenz, Nils Ellenfeld, Daniel Kant, Thomas Voss and numerous student assistants. Without their passionate participation and input, this work could not have been successfully conducted. Further my profound thanks goes to Tom Wytrwat for being the first reviewer of this manuscript.

Last but not least I also thank the following colleagues: Prof. Christoph Schick and Dr. Andreas Wurm of the Institute of Physics, University of Rostock for inviting me to their lab for measurements and valuable discussions; Dr. Philip Wenig and Dr. Andreas Klingberg of Lablicate GmbH for writing a modification of their open source chromatography software OpenChrom® just for my data, Anja Scholz and Astrid Poelders of the Institute of Environmental Technology and Energy Economics for GC measurements of pyrolysis gas and DSC; Reinhard Zschoche for revising the state of art chapter and finally the Central Laboratory for Chemical Analytics for many analyses.

Contents

1 Introduction

Despite major efforts in sustainability research, today's society still depends extensively on fossil resources. Energy and even more so chemicals are mainly generated from fossil feedstock. The disadvantage is not only that by using the resources massive amounts of greenhouse gases are emitted [1]. But also further detriments to the environment, human and animal health, like environmental mining issues or lung diseases from emissions, come along with their intensive use. Moreover, fossil fuel resources are limited. This limitation holds true despite the fact that due to increasing crude oil price and better exploitation technology the proven oil reserves have increased from 1980 to 2016 from about $640 \cdot 10^9$ barrels by the factor 2.5 [2]. The fossil fuel reserves-to-production ratio was estimated in 2015 to be about 50, 55, and 120 years for crude oil, natural gas, and coal, respectively [3]. Although the quantity of known resources has increased also the demand increases rapidly due to global population growth and rising living standards. In 2015 a global population of 7.3 billion people lived on earth. It is expected that with an annual growth of 1.18 % a population prospect of 8.5 billion people will be reached in 2030 [4]. Together with the growing population also living standards are rising. Thus, not only peak oil but also the peak of natural gas and coal are estimated within the next half-century [5]. A major share in the depletion of these resources has the production of petrochemicals. Approximately 5 % of annual oil and gas production is utilized in petrochemicals [6]. Moreover, the demand for and production of chemicals increases dramatically [7]. About $16.6 \cdot 10^6$ barrels of liquefied petroleum gas, ethane, and naphtha were utilized per day in 2015 and in prospect $21.2 \cdot 10^6$ barrels will be their daily demand for chemical production in 2040 [8]. The demand is further illustrated by the development of chemicals sales, which increased from $270 \cdot 10^6$ \$ in 1970 to $4200 \cdot 10^9$ \$ in 2010 and rise is estimated to accelerate to $6400 \cdot 10^9$ \$ in 2020 (not inflation-adjusted) [9].

Therefore, to meet the growing future demand, it is inevitable that the development of all regenerative energies, including bioenergy, are advanced to global prevalence. But, while the future energy demand might be met by regenerative energy technologies such as photovoltaic, thermal solar, hydrodynamic and wind power, only sustainable material cycles can be the solution to anthropogenic materials use. The future sustainable demand, replacing petrochemicals, can only be achieved by recycling, new products from biomass or possibly atmospheric CO_2. Unfortunately, already today the utilization of biofuels as an energy resource is expected to affect global food security [10]. Thus, to achieve the vision of regenerative materials cycles, while evading the competition with food production, other biomass sources than e.g. corn starch [11] or sugar cane [12] have to be evaluated [1].

This situation has of course not gone unnoticed by researchers, politicians and societies all over the world. Many countries have policies and roadmaps pushing the utilization and research of non-food-biomass for energy and chemicals. Intensive research is conducted on lignocellulosic biorefineries [1, 13–20]. Lignocellulosic biomass – a material composed of the carbohydrates cellulose and hemicellulose, and aromatic lignin – is the most abundant organic material on earth as it is the material grasses, straws, wood and other plant biomass are made of. Admitting that the technology of lignocellulosic biomass processing is of notably higher complexity [1], utilization of lignocellulosic biomass poses also a major chance as lignocellulose is abundant in huge amounts. For example, in the USA $400 \cdot 10^6$ t of straw (dry matter) remain unused, while in Germany $50 \cdot 10^6$ t could be used without diminishing natural field cultivation and organic fertilizing [15].

Pulping of lignocellulosic biomass is a well-established technology, while the production of bioethanol from the cellulose fraction is on the brink of commercial success. Regardless of the employed process, huge amounts of lignin are produced as a by-product. About 20 to 35 % of the utilized total biomass is comprised of lignin [21] and lignin amounts to approximately 20 % of the earth's land biosphere [22]. In industrial wood pulping about $80 \cdot 10^6$ t of lignin are produced [23] and with the increasing activity in lignocellulosic ethanol production, the available amount of lignin will further increase [22]. Common practice is the combustion of lignin for process energy, as its polyaromatic structure is very stable, and therefore, challenging to chemically modify [22]. But it is expected that modern biorefineries, will be energetically completely self-sufficient and thus have roughly 60 % of excess lignin [1].

Due to its interesting chemical structure, containing high fractions of aromatics, lignin could be a valuable source for aromatic products [20]. But due to the intrinsic natural resistance to microbial or enzymatic degradation (recalcitrance) it is tough to degrade biologically [1, 24]. Therefore, thermochemical conversion of lignin is an alternative as the harsh reaction conditions enable extensive degradation. Pyrolysis is such a process in which, under the absence of additional oxygen, the chemical structure is cleaved to form char, gas, and oil which is, in case of lignin, rich in aromatics. Possible products encompass biochar for soil enhancement, bio-bitumen, fuel additives, activated carbon, bio-resins, bioplastics, or specialty chemicals such as food additives and pharmaceuticals [20].

Many parameters influence the pyrolysis process performance. To achieve high liquid product yields, i.e. pyrolysis oil yield, a short oil vapor residence time in the hot reaction zone is necessary to prevent secondary cracking from oil to gas [25]. Furthermore, an excellent heat and mass transfer and high heating rates favor a high liquid product fraction. Circulating fluidized bed technology is characterized by exactly these attributes.

In a nutshell, circulating fluidized bed pyrolysis of lignin could be a promising process to reduce the humankind's profound dependence on fossil resources for the production of chemicals. CFB lignin pyrolysis does not only allow high yields of pyrolysis oil containing valuable chemical components, but also lignin is biosynthesized in large abundance.

2 State of art

This chapter provides an overview on the current technology for lignin pyrolysis. Because the biomass composition is directly impacting process performance as well as product composition, chemical composition and properties of lignocellulosic biomass are discussed, followed by a topical review on biomass pyrolysis technology with a particular emphasis on circulating fluidized bed reactors and lignin pyrolysis. Moreover, the parameters influencing pyrolysis performance are discussed and typical product yields, spectra, properties and applications of char, oil, and gas compared. The chapter is concluded with a discussion of the state of art in pyrolysis process modeling, including reaction pathways, kinetics, pyrolysis in fluidized bed reactors and pyrolysis refineries.

2.1 Lignocellulosic biomass

It is estimated that on annual basis about 1.7 to $2.0 \cdot 10^{11}$ t of terrestrial biomass is photosynthesized, with a fraction of $6 \cdot 10^9$ t used anthropogenically, mostly for food production and energy generation and only 3 to 3.5 % as material [15]. An additional sustainable biomass utilization potential of $1.3 \cdot 10^9$ t exists in the United States only [26, 27], of which roughly 95 % is lignocellulose. One example for lignocellulosic biomass is wheat straw, of which $529 \cdot 10^6$ t/a is generated worldwide. 43 %, 32 %, and 15 % are produced in Asia, Europe, and North America, respectively. [14] It is a very attractive resource for the production of bioethanol because of its fast growth, low cost and low lignin content [28, 29]. In Germany $50 \cdot 10^6$ t straw (dry matter) is available for production of chemicals and fuel without negatively affecting the agricultural nutrient cycles [15].

The cell wall of vascular plants mostly consists of the biopolymers: cellulose, hemicellulose (polyoses), and lignin [14, 30, 31], and is hence referred to as lignocellulosic biomass. The composition depends on biomass type and origin (cf. Table 2.1). Additional biopolymers in lignocellulosic biomass are polyhydroxy fatty acid esters, polyisopropenoids (e.g. terpenes and steroids), glycosidic pectins and energy storing carbohydrates (sugars and starch). Other minor constituents can be dyes, pigments, flavors, alkaloids, and inorganic matter, with their amount and composition depending on the biomass type. [14] The biopolymers are heavily intertwined with each other [30] but can be separated by pulping technologies [30–32]. Cellulose is a valuable resource for pulp and paper industry as well as for production of renewable ethanol [33], levoglucosan (potentially useful as polyol) [34] and others. From hemicellulose furfural based nylon can be produced [14]. Lignocellulosic biomass consists of 6 to 33 wt.-% lignin [14, 31], which is the largest renewable resource of aromatics on the planet [30]. In conclusion, lignocellulosic biomass such as straw and forest resources is available in great abundance and a renewable resource for both fuel and chemical production.

Table 2.1: Composition (wt.-% dry matter) of biomass types (data from [14, 31])

biomass type	cellulose	hemicellulose		lignin
		hexoses	pentoses	
hardwood	30-43	2-5	17-25	18-25
softwood	40-48	12-15	7-10	26-33
straw/ grasses	35-41	0-5	15-28	6-24

2.1.1 Cellulose and hemicellulose

Cellulose

Cellulose and hemicellulose fibers provide structural support to plant cells. Cellulose, as the main constituent of the cell wall, is the most prevalent biopolymer. [14] It is approximated that the annual biosynthesized yield of cellulose is $1.3 \cdot 10^9$ to $1 \cdot 10^{11}$ t/a [14, 35]. Cellulose is a non-branched water-insoluble polysaccharide built from glucose monomers linked by β–1,4–glycosidic bonds yielding a syndiotactic β–1,4–polyacetal of cellobiose (4–O–β–D–glucopyranosyl–D–glucose) (cf. Figure 2.1) [14, 33, 35]. The length of a cellulose chain ranges from several hundred to tens of thousands of β–glucose molecules [14]. Its basic structure can thus be expressed by multiples of cellulose $C_6 PH_{10 P+2}O_{5 P+1} \approx (C_6H_{10}O_5)_P$, where P is the degree of polymerization (number of glucose units with $M_{glucose} = 162$ g/mol) [14, 35]. The degree of polymerization for technical (treated) cellulose is in the range of about 1000 to 3000 [23] and for untreated cellulose in the range of 5000 to 10 000 [36]. The molar mass of a cellulose polymer depends on its degree of polymerization: $M_{cell} = M_{glucose} \cdot P + 18 \approx 162$ g/mol $\cdot P$ [35]. Cellulose is a nonmelting polymer [35].

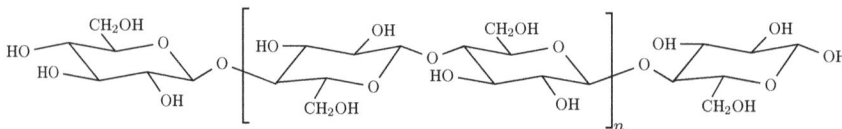

Figure 2.1: Model chain structure of cellulose

Hemicellulose

Hemicellulose (Greek: hemisys=half) is an amorphous polyose (cf. Figure 2.2) in the plant cell membrane and consists of hexose and pentose subunits, among them are D–glucose, D–galactose and D–mannose (hexoses) and D–xylose and L–arabinose (pentoses). [14, 33] Furthermore, hemicellulose can encompass sugar acids (uronic acids) such as D–glucuronic, D–galacturonic and methylgalacturonic acids. The bonding frequency and composition (amount of specific hexoses and pentoses) in hemicellulose depends on the biomass type and its source. The main linkages are xylan β–1,4–linkages. In hardwood, more xylans are found than in softwood, which contains more glucomannans than hardwood. [33] The degree of polymerization of technical hemicellulose is within 50 to 200, and thus considerably lower than that of cellulose [23, 36].

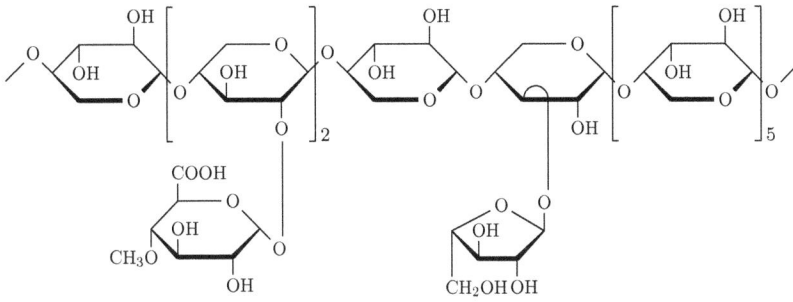

Figure 2.2: Model structure of hemicellulose (arabino–4–O–methyl–glucuronoxylan), as proposed by [37–39] in [14]

2.1.2 Lignin

Lignin, which comprises about 20 % of the planet's biosphere mass [22], is after cellulose the second most common organic substance in the world [14]. It was first named in 1813 by the Swiss botanist de Candolle, M. A. P. [40] after the Latin word lignum (wood). Together with hemicelluloses, lignin is a binding agent between adjacent cells and is heavily intertwined with hemicellulose and cellulose microfibrils in the cells itself, forming a composite structure with outstanding strength and elasticity. Furthermore, lipophilic lignin controls water permeation as well as nutrient and metabolite transport. Finally, lignin defends plants against microorganisms by inhibiting penetration of harmful enzymes into the cell walls. [30, 31]

Lignin is mostly obtained as a by-product of pulping processes. These processes are Kraft, sulfite, soda, organosolv, and aquasolve pulping and other processes of minor importance [13, 30–32, 41]. The most important process is Kraft pulping with an annual global production of 73 %, which relates to $170 \cdot 10^6$ t pulp [32]. The majority of lignin is burnt for pulping chemical recovery and supply of pulping process energy; less than 2 to 5 % is isolated and sold [22, 31]. Recent research interest in alternative pulping processes has increased due to intensified work on lignocellulosic biorefineries [6, 17]. The aim is to sustainably produce bioethanol as fuel and lignin as a by-product.

The content of lignin in biomass increases from grasses/straw over hardwood to softwood (Table 2.1). Lignin is an amorphous randomly cross-linked polyphenol, consisting of the phenylpropane (C_9) subunits coumaryl, coniferyl, and sinapyl alcohol, which have zero, one, and two methoxyl groups (Figure 2.3), respectively [14, 30, 31, 42]. Depending on plant type and species the contents of these primary monomer units vary (cf. Table 2.2) [30]. Softwood lignin is primarily built from coniferyl alcohol (guaiacyl structure) and thus called G-lignin. Hardwoods, on the other hand, contain almost equal amounts of coniferyl and sinapyl alcohol (syringyl structure) and therefore their lignin is categorized as GS-lignin. [30, 31] Grass and straw lignin contains all of the three monolignols (additionally the coumaryl structure p-hydroxyphenyl) and is thus abbreviated (GSH-lignin) [29–31]. Diverse functional groups can be found in the lignin structure. Predominantly these groups are methoxy, phenolic, and aliphatic hydroxy, benzyl alcohol, non-cyclic benzyl ether, and carbonyl groups [30]. Reported frequencies per C_6C_3 unit in spruce milled wood lignin are 1.09 for aliphatic OH, 0.26 for phenolic OH, and 0.2 for total carbonyl [43].

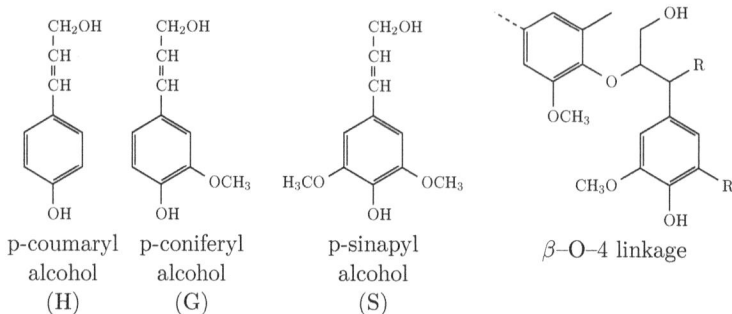

Figure 2.3: Aromatic phenylpropane subunits of lignin and most frequent interunit linkage

A multitude of different linkages irregularly connect the various primary monomer unit's allyl ethers to form a network with O–O, ether, and C–C bonds. Depending on the biomass type the proportion of these linkages varies. Softwood and hardwood are connected by ether-moieties in more than two-thirds of the linkages. [14, 30, 43] The most common linkage is β–O–4–aryl ether (cf. Figure 2.3) with e.g. 60 % in birch (hardwood) and 46 % in spruce (softwood). Other important linkages include β–5–phenylcoumaran, α–O–4–, 4–O–5–diaryl ether, β–β–resinol, 5–5–biphenyl and β–1–1,2–diarylpropane motifs [14, 30, 42, 43]. The proportions for birch and spruce wood are given in Table 2.3.

Table 2.2: Lignin composition of phenyl-propane units [30]

| biomass type | phenylpropane unit, % | | |
	coumaryl (H)	coniferyl (G)	sinapyl (S)
hardwood	-	50	50
softwood	-	90[†]-95	5-10
grasses	5	75	25
wheat straw[‡]	5	49	46

[†]from [43], [‡]from [29]

Table 2.3: Interunit linkages of wood lignin [43]

| linkage type | proportion in % | |
	softwood (spruce)	hardwood (birch)
β–O–4	46	60
α–O–4	6-8	6–8
4–O–5	3.5-4	6.5
β–5	9-12	6
β–1	7	7
β–β	2	3
5–5	9.5-11	4.5

Molar mass

The molar mass of isolated lignins depends on the pulping conditions and biomass type [44]. Most isolated lignin, e.g. from the Kraft, soda, and sulfite pulping have molar masses in a range between 3000 and 20 000 g/mol. [30, 31] An example is wheat straw lignin from an organolsolv process with $M_w = 3960$ g/mol and $M_n = 2330$ g/mol [45]. The M_w values for Kraft lignins are generally below 10 000 g/mol [31].

Glass transition

The lignin glass transition temperature ϑ_g depends on moisture content, raw material, production process, molar mass, and measurement procedure. Hardwood lignins have ϑ_g values between 65 and 85 °C which is lower than in softwoods (90 to 105 °C). For isolated lignins the following ϑ_g values have been found: MWL lignins: 110 to 180 °C [31, 46] and

Kraft lignins: 102 to 174 °C. Moisture has a severe effect: values of 195 °C when dry and 90 °C when containing 27 wt.-% moisture have been reported. The effect of lignin molar mass is even more severe, i.e. a ϑ_g of only 32 °C at low molar mass ($M = 620\,\text{g/mol}$) and a ϑ_g of 173 °C at the highest molar mass $M = 180\,000\,\text{g/mol}$. [31, 47] Furthermore, Nowakowski et al. [48] reported softening of amorphous lignin between 120 to 180 °C and Hatakeyama et al. [49] a ϑ_g for dioxane lignin at 122 °C.

2.2 Biomass pyrolysis

Pyrolysis is a thermal decomposition process for organic materials. As opposed to combustion or gasification no oxidation by oxygen (or an oxygen donor) occurs. Pyrolysis, in the absence of air, is utilized to crack the feedstock to gas, char, and oil. The process conditions substantially affect the product composition. At low heating rates < 200 K/s, moderate temperatures < 400 °C, and long solids residence times (up to several days) mostly char is produced [36]. This process, called conventional or slow pyrolysis, is applied since centuries to produce e.g. charcoal. A more recent technology is fast pyrolysis, also called rapid or flash pyrolysis. Fast pyrolysis at higher temperatures between 500 and 1000 °C, short vapor residence times < 2 s, and high heating rates of 1000 to 10 000 K/s [50–53] aims at producing high amounts of liquid product for energetic or chemical valorization [36]. The vapor residence time for high liquid yield pyrolysis has to be kept short (typically less than 2 s) to prevent secondary degradation of liquid to gaseous compounds [54].

A typical layout [55–57] of a biomass pyrolysis process is shown in Figure 2.4. The biomass is fed into the pyrolysis reactor where it is cracked to oil, gas, and char. Downstream, first solids and then permanent gases are separated from the product oil. Separation of the solids is necessary because they promote secondary cracking of pyrolysis oil [25], pyrolysis oil aging, oil instability, and other application barriers. In most cases, cyclones are used as they are easily maintained high-performance separators and downstream separation of solids from the liquid phase proves difficult [53]. Additionally, candle filters, granular filters, and others are investigated for pyrolysis process application [58–62]. To achieve high liquid yields, the oil vapors are cooled rapidly (quenched) to prevent secondary reactions from oil to gas. Optionally, the by-products (permanent gas and char) can be combusted to supply the endothermic pyrolysis energy demand. Furthermore, electrostatic precipitators are widely used to separate aerosols from pyrolysis gas and vapors.

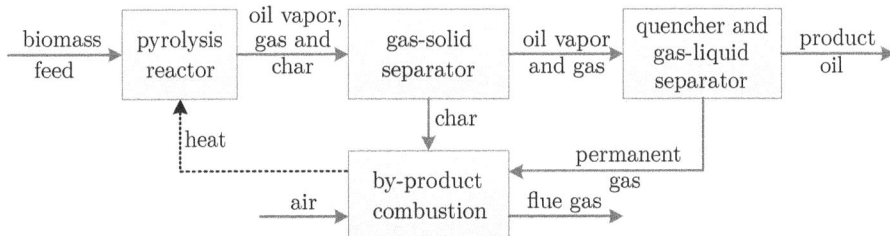

Figure 2.4: General scheme of biomass pyrolysis system

A large variety of reactor types for fast pyrolysis has been investigated mainly at laboratory and pilot scale: ablative reactors in which the feedstock is heated by contact with a hot surface [54, 56, 63] like cyclone reactors [64–68], vortex reactors [54, 69–72], auger or double screw reactors [56, 73, 74], rotating cones, rotary kilns, and hearth furnaces [75–77]. Furthermore, reactors with mainly convective and conductive heat transfer such as fixed bed reactors [78, 79], fluidized beds such as (conical) spouted beds [52, 80–82], bubbling/ stationary fluidized beds [51, 53, 55, 56, 63, 83, 84], fluidized beds with mechanical fluidization [85–87], and circulating fluidized bed reactors [88–93] have been applied to pyrolysis. Additionally, reactors that transport heat by means of radiation and convection: entrained flow reactors [94–96] and microwave pyrolysis reactors [97–99] have been used. Lastly, especially thermogravimetric analysis and tubular reactors coupled with various online product analysis systems are widely applied for analytical investigation of pyrolysis and pyrolysis mechanisms [100–104]. Besides the analytical equipment, of the above reactor types, fluidized beds are used most frequently [54, 105].

2.2.1 Pyrolysis in circulating fluidized bed reactors

Circulating fluidized beds (CFB) have the advantage of short vapor residence time and good heat transfer (about 80 % conduction, 19 % convection, 1 % radiation [54]) leading to high reaction rates. An additional advantage is that CFBs are potentially suitable for larger throughputs, as CFB technology is widely used at very high throughputs in the petroleum and petrochemical industry [56, 106]. Summarized advantages and disadvantages are [54–56, 70]:

+ good mixing

+ good temperature control

+ high heat transfer & reaction rates

+ short vapor residence time

+ large feedstock particle size possible (up to 6 mm)

+ can be coupled with a char combustion reactor for heat supply

+ high throughput possible ($\sim$ 60 t/d and bigger)

+ catalytic bed material can be used

− possible liquid cracking by hot solids

− char attrition and breakage leading to higher solids content in pyrolysis oil

− possible attrition and breakage of catalytic bed material leading to higher process costs

− increased complexity (compared to BFB)

A typical layout [55, 56] of a biomass pyrolysis process with circulating fluidized bed reactor is shown in Figure 2.5. It shows an integrated system with a CFB pyrolysis reactor, which is coupled with a stationary fluidized bed for char combustion. The CFB consists of a pyrolysis riser reactor, a cyclone for gas-solid separation, a connection to the char combustion reactor and a return leg into the riser. Two loop seals are necessary to prevent gas exchange between the reactors. The recirculating bed material is heated by char combustion, supplying the heat demand for pyrolysis. In this example, secondary solid separation (e.g. additional cyclones) is followed by a scrubber for fast quenching of the pyrolysis reaction and an electrostatic precipitator. The non-condensable permanent gas can partly be recycled to supply the fluidization gas for the pyrolysis reactor.

In 1999, ENSYN was the only commercially operating organization giving a performance guarantee for their fast pyrolysis plants [53]. The reactor is a common CFB riser with

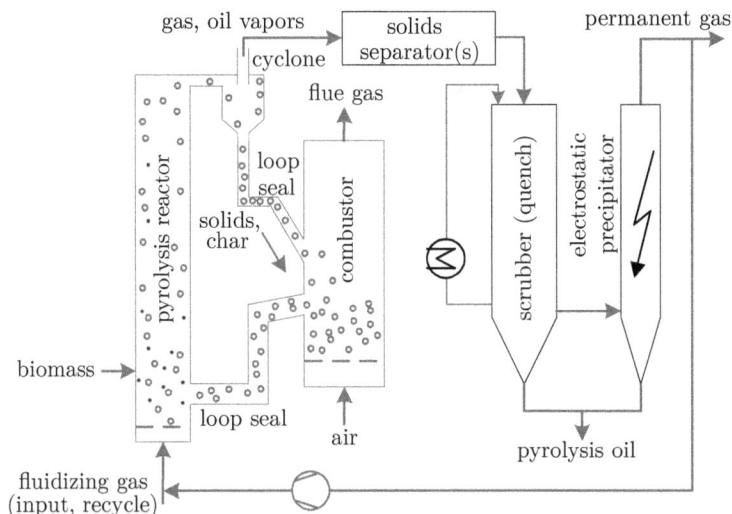

Figure 2.5: Circulating fluidized bed system for biomass pyrolysis with integrated char combustion and product oil separation

syphon and return leg [107–110]. Further examples of plants include a 36 t/d unit at Red Arrow, WI, USA and some plants in research & development organizations (e.g. ENEL, Italy and VTT/Metso/UPM, Finland) [53, 56]. Today, Envergent Technologies (a joint venture of ENSYN and Honeywell) has developed the fast pyrolysis plant with the largest liquid product capacity of about 30 MW [111], although for the produced biofuel no significant market has been established yet [111]. In 2007, another CFB reactor system was investigated. The biomass is fed into the bottom of the riser section. The entrained char is returned to a bubbling fluidized bed at the bottom of the riser, where it is partly oxidized for supply of pyrolysis energy. [88, 89]. A recent 2013 comprehensive review of research and commercial activities in the field is given by Meier et al. [112], but until today no in-depth investigation of lignin pyrolysis in a CFB system has been reported.

2.2.2 Lignin pyrolysis

Lignin can be obtained from various biomass sources by diverse pulping processes. Therefore, a large variety of lignins has been investigated as pyrolysis feedstock. Examples are: wheat straw lignin [113], Kraft lignin [47, 114–116], alcell lignin [117], organosolv lignin (from softwood and wheat straw) [29, 48], soda lignin (from grasses and wheat straw) [48], bamboo lignin [118], alkali lignin [119], lignin from corn cob acid hydrolysis residue [120], and milled wood lignin (MWL) [121]. Next to the lignin variety also different reactors for lignin pyrolysis have been investigated. Analytical methods and reactors range from pyrolysis-GC-MS [47, 48, 120], TG-FTIR [120, 122], TGA [48, 119], pyroprobe [118, 123], heated screen [119] over tubular laboratory scale reactors [102, 116], microwave pyrolysis [97] and centrifuge reactors [124] to fixed bed [115, 117], and fluidized bed reactors [29, 48, 85–87, 125].

Lignin pyrolysis – especially in fluidized beds – has been reported to be difficult by many investigators [29, 48, 87]. Nowakowski et al. [48] carried out round robin pyrolysis of a soda and an organosolv lignin in several laboratories and reactors. They found that the pyrolysis of the soda lignin (94 wt.-% lignin content) led to problems already when feeding pneumatically or by auger feeding into the reactor. It melted in the pneumatic feeding lance and the screw feeder, respectively. Additionally, molten lignin in the fluidized bed resulted in bed agglomeration and subsequent defluidization. The authors conclude that fluidized bed pyrolysis of this lignin is nearly impossible. The same melting behavior leading to bed agglomeration and defluidization was observed for wheat straw lignin (lignin purity > 90 %) pyrolysis at 550 °C. Removing the lignin fines prior to feeding resolved the problem [29]. For the softwood organosolv lignin (about 40 wt.-% carbohydrate content, estimated from elemental analysis) agglomeration was also observed but only in one lab and to less extent [48]. Due to the melting at 150 to 200 °C, which together with slow reaction induces a liquid phase and therefore bed agglomeration and defluidization [87], researchers at ICFAR, University of Western Ontario developed a mechanically agitated fluidized bed reactor with an internal mechanical stirrer [85–87]. For further reading on recent lignin pyrolysis research activities, the work of Lago [85] is suggested.

2.2.3 Influence parameters

2.2.3.1 Main parameters

The yield and composition of pyrolysis products can be adjusted by careful selection of process parameters. The main parameters are temperature, vapor and solids residence time as well as heating rate. Residence time and heating rate are primarily defined by reactor type; nevertheless, they can be adjusted within the limits of a given system.

Temperature

The main control parameter is the pyrolysis temperature. Numerous studies on the influence of this parameter on product yields have been conducted [51, 56, 119, 126–128]. Virtually all feedstock and reactor variations have been studied. Generally, the gas yield increases with rising temperature, because more feedstock is converted into volatile products and secondary reactions convert more pyrolysis oil to gas [51, 56, 84, 126]. Complementarily, the char yield decreases continuously [51, 56, 84, 126], while the oil yield increases until an optimum temperature is reached [51, 56, 84, 119, 126–128]. Beyond this optimum, the secondary cracking reactions from oil to gas become more substantial [51, 56, 84, 119, 126–128]. For illustration, two examples for fluidized bed pyrolysis are given in Fig. 2.6.

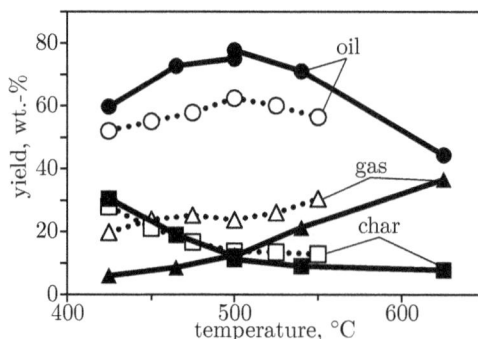

Figure 2.6: Fluidized bed pyrolysis yields of sawdust from wood mixture [128] (empty symbols) and hybrid poplar aspen wood [84] (filled symbols)

Heating rate

Hoekstra et al. [129] found that the heating rate severely impacts the liquid yield for cellulose and lignin pyrolysis in a laboratory scale wire mesh reactor. The liquid yield from cellulose and lignin increased by 6 wt.-% and 24 wt.-% respectively, if the heating rate was increased from 50 to 6000 K/s. This increase occurs at the expense of both gas and char yield. For lignin, the oil's molar weight distribution was narrower at the lower heating rate of 50 K/s, which could be explained by a longer vapor residence time in the hot zone. The decrease of char yield with increasing heating rate is confirmed by Chhiti et al. [130]. Because it is difficult to measure the particle heating rate in fluidized beds accurately, it is usually approximated by simulations. Di Blasi [131] investigated the heating rates of cellulose particles in fluidized beds by simulation and found that the heating rate depends on particle size and conversion. The heating rates estimated vary from about 8 to 1100 K/s for maxima (low conversions), but on average (high conversion) the actual values reduce to 1 to 300 K/s. Kersten et al. [132] calculated an average rate of 50 K/s for heating a 2 mm particle injected into a fluidized bed from ambient temperature to 490 °C. The influence of biomass particle size on the heating rate in a fluidized bed was approximated by Salehi et al. [128]. Particles smaller than 590 µm exhibit a heating rate of $\geq$ 2200 K/s. For larger particles in the size range of 1000 to 1400 µm, the heating rate first decreases sharply before it reaches about 250 K/s. Further estimations for heating rates at fast pyrolysis conditions are: fluidized bed 1000 to 10^5 K/s [50, 51] and entrained flow up to 10^4 K/s [133]. It should be mentioned here that these latter high values are mostly initial heating rates (zero conversion) for very small particles, neglecting the energy consumption by pyrolysis [134].

Vapor residence time

For achieving a high gas yield a long vapor residence time is essential. A high liquid yield, on the other hand, is favored by short vapor residence times, suppressing secondary reactions converting oil to gas, especially at elevated temperatures. Below 350 °C these reactions are slow [25]. An example is the pyrolysis of aspen wood in a fluidized bed reactor. It was shown that with an increase of vapor residence time from 0.4 to 1.1 s the liquid yield decreased from 57.5 wt.-% by about 12 wt.-% [51].

2.2.3.2 Other parameters

Next to the above-mentioned main parameters (temperature, heating rate, and vapor residence times), further parameters must be considered, such as feed mass flow, biomass particle size, gas atmosphere (e.g. steam, inert gas like nitrogen, or recycled pyrolysis gas), pressure, and feedstock mineral matter content. Except for mineral matter and gas atmosphere, these parameters are closely related and their influence can in most instances be attributed to the main parameters. For instance, the feed mass flow primarily influences the heating rate and residence times.

Biomass feeding rate

Park et al. [135] investigated the pyrolysis of sawdust in a bubbling fluidized bed. As the variation of mass flow was not decisive and the mixing behavior and heat transfer in such a system can be considered optimal, it can be assumed that only the residence time of the gaseous products was affected. The authors showed that the char yield stayed constant, while the oil yield increased to the same degree as gas yield decreased, likely a result of decreased vapor residence time with increasing mass flow. However, changing the pyrolysis system feeding rate to a larger degree will also have an effect on the heating rate.

Biomass particle size

The biomass particle size affects the heating rate, pyrolysis particle temperature as well as residence times (e.g. in fluidized beds with considerable entrainment and segregation effects). A larger particle will heat up more slowly and develop a temperature gradient. Therefore, the actual pyrolysis temperature inside the particle will be lower than the reactor temperature, resulting in a different product spectrum. Furthermore, a larger particle impairs the removal of gaseous products from the reaction zone, as the gas must diffuse through the particle itself, resulting in slower mass transfer compared to smaller particle size. The particle matrix itself might catalyze the decomposition of primary volatiles. This effect can be explained by a sequence of deposition, adsorption and polymerization of primary volatiles on the solid pore surface [136]. Due to these correlations with other parameters, it is not straightforward to predict the effects of changing particle size. Investigations in fluidized and fixed beds showed that with increasing particle diameter char yield can vary from decreasing [51] to unaffected [135, 137] and slightly increasing [128]. Likewise, for gas and oil, intangible trends are reported in the literature. The trends vary from increasing [51, 128] to constant [137] and decreasing [135] and from an optimum [135] to constant [137] and globally decreasing [128, 138] with increasing particle size. To achieve high heating rates usually finely ground feedstock is required [54].

Pressure

Another parameter is the reactor pressure [34, 77, 139–143]. By applying reduced pressure, the vapor residence time in the hot zone is decreased [144] (and thus secondary reactions suppressed). An example is the pyrolysis of lignin in N_2 at $600\,K/s$, $650\,°C$ and $0.000\,133\,bar$. At these conditions, the oil yield with $58\,wt.-\%$ was found to be $16\,wt.-\%$ higher than at atmospheric pressure [119]. On the other hand, the oil yield from pinewood pyrolysis in a conical spouted bed did not change considerably at vacuum conditions [52]. This finding might be explained by the already sufficient mass transport at ambient pressure in such a reactor system, i.e. optimal mixing, high dilution of the hot vapors with fluidization gas and short vapor residence times [144]. Cellulose pyrolysis was investigated in a pressure range of 1 to 5 bar [145, 146]. The experiments in the laboratory scale steam flow reactor revealed that the specific surface area of cellulose char decreases with increasing pressure. Furthermore, the elevated pressure resulted in a decreased oil and increased gas yield, which might be explained by the higher vapor pressure of the oil components leading to intensified crosslinking reactions with a release of gases.

Reactor atmosphere

Diverse studies on the effect of different gases on pyrolysis yields have been carried out. Investigated were reductive and mild oxidative atmospheres. Reductive gases analyzed are CO and H_2. The mild oxidative gases used are CO_2 as well as CH_4 and H_2O. For steam, conflicting study results exist. Some researchers found steam to have a negligible effect on pyrolysis below $800\,°C$ [136], whereas others measured small to significant impacts on pyrolysis yields and product composition [137, 147–153]. Several effects have been linked to steam, i.e. an enhanced tar evaporation and desorption [147, 151], hydrogen donation [147, 151], mild oxidation at temperatures of 800 to $1200\,K$ [151], decarboxylation, and decarbonylation (release of $-COOH$ and CO, respectively) [151], as well as an improved heat transfer [147]. The char yield decreased with the addition of steam [150, 153], while oil yield increased [137, 150, 152]. Analogously to coal pyrolysis at elevated pressure and in the presence of steam [154, 155], cellulose pyrolysis was investigated in a pressure range of 1 to 5 bar [145, 146]. The experiments in the laboratory scale steam flow reactor revealed that cellulose char from steam pyrolysis exhibits a larger surface area than char from inert nitrogen pyrolysis. The influence of the other gases can be read elsewhere [42, 119, 136, 156–160].

Mineral matter

Various studies revealed that the mineral content of biomass influences the pyrolysis product formation [161]. Inorganics in biomass are alkali and alkaline earth metals Li, Na, Mg, K, Ca, Ba, various metals Al, Ti, V, Cr, Mn, Fe, Ni, Zn, and others Si, P, S, Cl, etc. The inorganics are associated with counter ions or connected to organic acids. [162, 163] Popular methods for investigating the influence of minerals on pyrolysis product distribution are solvent washing of biomass, with subsequent salt addition or ion exchange [164]. The difference is that salt addition has compared to ion exchange a smaller but still major effect on biomass decomposition [164]. Fahmi et al. [165] investigated the liquid yield of various biomasses (mostly grasses, washed and untreated) with an ash content of 0 to 8 wt.-%. With increasing ash content, the oil yield decreased from about 72 to 45 wt.-% whereas the char yield increased. This insight is supported by Eom et al. [166], who pyrolyzed rice straw (both untreated and demineralized) in a fluidized bed. Furthermore, the addition of potassium acetate to short rotation willow coppice and a synthetic mixture of cellulose, hemicellulose, and lignin increases the char yield [167]. The addition of $Fe(NO_3)_3$, $Al(NO_3)_3$, $Ca(NO_3)_2$, K_2CO_3, $Mg(NO_3)_2$, and Na_2CO_3 to demineralized pine bark with successive thermogravimetric analysis (TGA) revealed an increase in char yield [168]. The strongest effects were found for $Al(NO_3)_3$, Na_2CO_3, and K_2CO_3 in decreasing order. Na_2CO_3 added to tobacco stalk and yellow pine wood on the other hand slightly improved the oil yield, particularly with respect to methanol, acetic acid and 1–hydroxy–2–propane yield [169]. Cation exchange in cottonwood with K^+, Li^+ and Ca^{2+} has an increasing and decreasing impact on char and tar yield in vacuum pyrolysis, respectively [164]. Moreover, potassium catalyzes pyrolytic reactions resulting in more CO_2, CO and formic acid being formed from polysaccharides and acetic acid from hemicellulose [170]. The effect of K, Mg, and Ca on degradation products by addition of their chlorides to demineralized poplar wood xylem tissue was investigated by Eom et al. [171]. The char yield increased from 10.5 to 19.6 wt.-% for potassium, whereas magnesium and calcium had a minor effect. Furthermore, K-addition led to an increase of acetol, butanedial, cyclopentenes, phenol, guaiacol and syringol whereas the yield of levoglucosan was decreased. Magnesium, on the other hand, increases the levoglucosan yield. This finding is supported by Richards and Zheng [164], who found K, Li, and Ca to decrease the yield of levoglucosan, whereas other ions like transition metals tend to increase the levoglucosan yield. Removing almost the entire mineral content by washing poplar wood or stake cellulose can increase the levoglucosan yield while decreasing hydroxyacetaldehyde yield [172, 173]. Further studies evidence the influence of Ca^{2+} [174, 175], Fe^{3+} [176], acidic, alkaline and neutral salts [177] on product yields.

Mineral matter has an impact on pyrolysis product distribution for lignin as well. For example, sulfur in form of sulphoxide or sulphone, contained in Kraft lignin, facilitates lignin depolymerization [42]. The presence of cations Na^+, NH_4^+, and Ca^{2+} influences lignin decomposition. The strongest effect in lignin pyrolysis was observed for Na^+, which increases char and gas yield at the expense of oil yield [178]. Jakab et al. [179] investigated the influence of sodium addition to MWL. Sodium was found to facilitate char formation and cleavage of functional groups while the volatiles yield decreases. Sodium enhances demethoxylation (cleavage of $R-OCH_3$) rather than demethylation (cleavage of $R-CH_3$), dehydration and decarboxylation are promoted, while C-O bond scission is preferred, i.e. sodium-catalyzed heterolytic reactions. Furthermore, formaldehyde and CO formation are suppressed [179, 180]. Sodium contained in the lignin side groups $-COONa$ and

−CONa increases the yield of alcohols and hydrocarbons, whereas the yield of benzenes is reduced [181]. The impact of K_2CO_3 and $ZnCl_2$ on pyrolysis behavior of cellulose, xylan, and lignin was studied by Rutkowski [182]. All combinations except lignin with K_2CO_3-addition showed a constant or decreasing oil yield. The lignin oil yield with K_2CO_3-addition increased as K_2CO_3 favored dehydration and demethoxylation such that the aliphatic hydrocarbons yield increased whereas oxygen containing component yields of monocyclic aromatic hydrocarbons, phenols and carboxyl group containing substances decreased. Treatment of MWL with NaCl and $ZnCl_2$ was carried out by Jakab et al. [183]. These salts lead to a char yield increase by 4 and 6 wt.-%, respectively. NaCl further decreases CH_2O yield whereas H_2O and CH_3OH increase. The Lewis acid ZnCl2 catalyzes ionic reactions, i.e. heterocyclic cleavage of C−O and C−C. Also, more CH_3OH compared to the untreated but less than for NaCl-supplemented MWL is formed. Furthermore, potassium is known to increase the yield of CH_3OH from lignin [170].

2.2.4 Product yields, spectrum, properties, and applications

2.2.4.1 Pyrolysis oil

The pyrolysis oil yield ranges from 40 to 95 wt.-% [36, 102] and sometimes even as low as 12 wt.-% [48], depending on pyrolysis feedstock, i.e. cellulose, hemicellulose and lignin content, inorganics content as well as pyrolysis conditions such as heating rate, pyrolysis temperature and vapor residence time. High lignin content leads to a lower liquid yield [36]. Qu et al. [102] found maximum oil yields of 60 wt.-%, 53 wt.-%, and 40 wt.-% at 400 °C, 450 °C, and 500 °C, respectively, when pyrolyzing cellulose, hemicellulose and lignin in a fast tubular laboratory reactor. A study on the pyrolysis of cellulose in a fluidized bed [184] also showed a maximum oil yield in the magnitude of 60 wt.-%. The tar yield of Kraft and hardwood Alcell lignin in a fixed bed yielded 20 wt.-% and 25 wt.-%, respectively [115, 117]. The pyrolysis of wheat straw lignin in a centrifuge reactor at 500 to 550 °C, a feed of 340 g/h and a gas residence time of 0.8 s yielded about 32 wt.-% organic oil and 10 wt.-% reaction water [124, 163]. Furthermore, oil yields for lignin pyrolysis in fluidized beds were reported in the range of 16 to 57.7 wt.-% with the optimum mostly in the temperature range of 450 to 550 °C] [29, 48, 125]. For entrained flow pyrolysis at 700 °C 11.7 to 36.6 wt.-% [48] and mechanically fluidized bed pyrolysis at about 550 °C 45 wt.-% [85] were found.

Biomass pyrolysis oil is a complex mixture of several hundred organic components as well as inorganic constituents. The main compounds found in biomass-derived pyrolysis oils are: acids (mostly carboxylics), alcohols, aldehydes, alkanes, alkenes, esters, furans, ketones, nitrogen compounds, miscellaneous oxygenates, phenolics, syringols, catechols, guaiacols, other aromatics, carbohydrates (sugars), and water [36, 47, 48, 101, 102, 113, 114, 123, 156, 162, 185, 186]. Specific examples are acetic acid, furfural, furan, 1-Hydroxy-2-propanone, and water for hemicellulose and levoglucosan, 5-hydroxymethylfurfural, formaldehyde, acetic acid, acetol, and water for cellulose [101, 185]. Lignin yields small quantities of monomeric phenolics (such as phenol, cresol, guaiacols, syringols), acetic acid, CH_3OH, formaldehyde, acetaldehyde, benzene, water and furthermore, mainly from lignin, but also from cellulose a large amount of oligomeric oil species with molar mass ranging from few hundred to several thousand g/mol [36, 47, 48, 113, 114, 123, 162, 185, 186]. Part of the oligomeric products from lignin is so-called pyrolytic lignin, which has on average a higher molar mass than the water-soluble pyrolysis oil fraction and can be precipitated from the oil by the addition of deionized water [46, 187, 188]. The pyrolytic lignin from MWL pyrolysis in a fluidized bed was found to have mostly light brownish color and yielded 8 to 25 wt.-% of lignin feed [46, 188]. Additionally, extractives produce waxy bio-oil species like fatty acids and rosin acid [185]. Due to its complex composition and high reactivity the pyrolysis oil is not distillable and highly inhomogeneous, which can lead to phase separation [56]. The latter is also caused by water in the oil, which can for various biomasses contribute to pyrolysis oil in a content of 5 to 55 wt.-% [46, 54, 70, 128, 142]. Inorganic species present in the biomass are also found in the pyrolysis oil. The single mineral species content is mostly in the range of 0 to 20 mg/kg (rarely in the magnitude of 100 mg/kg) and thus 10 to 10000 times lower than the content in char [189]. The total solids content (inorganics and char) of pyrolysis oil is typically found to range from <0.1 to 0.4 wt.-% [36].

Pyrolysis oil has a dynamic viscosity of 5 to 350 mPa · s [54, 185, 190] and a kinematic viscosity of $1.1 \cdot 10^{-5}$ to $9.7 \cdot 10^{-5}$ m^2/s [46, 70], which depends on biomass feedstock, water, alcohol, and inorganics content as well as storage time [190, 191]. The acids contained in the oil leads to a pH between 2 and 5.5 [36, 46, 54, 190]. Pyrolysis oil densities are in the range between 0.9 to 1.3 g/cm^3 [36, 54]. The higher heating value of pyrolysis oil ranges from 13.9 to 41 MJ/kg. This range comprises raw pyrolysis oil with high water content (low heating value) and water free oils with high heating value. Due to its mostly high water content, raw pyrolysis bio-oil typically has a higher heating value (HHV) at the lower end of the range, which is considerably lower than the HHV of petroleum with about 40 MJ/kg [192]. For example, Raveendran and Ganesh [193] summarized the HHV of raw pyrolysis oil from diverse lignocellulosic feedstocks to be in the range of about 22 to 25 MJ/kg.

Several application prospects exist for pyrolysis oil. It can be used as combustion or transportation fuel, for power generation and the production of chemicals and resins from either pyrolysis oil fractions or single components. [20, 162] As liquid products are easier to handle and may be stored and transported, and thus can be used offside the production site, it is beneficial to use pyrolysis oil as combustion or transportation fuel or fuel additive substitution in fossil fuels [20, 53, 162, 194]. The use of pyrolysis oil in power generation was successful as it can be combusted in diesel engines and gas turbines [53, 54, 162, 194]. Problems with high viscosity and suspended char are benign, but processes have been run only demonstration scale [53] so far. Fractions of pyrolysis oil are used as binder for pelletizing and briquetting combustible organic waste materials [162], heavy pyrolysis oil tar for roofing or roads [195], calcium salts of carboxylic acids as environmentally friendly road de-icers [54, 194], polyphenols or pyrolytic lignin to substitute phenol in phenol-formaldehyde resins [7, 54, 194, 196–198], and in the production of adhesives, especially for fiberboard production [7, 162]. Furthermore, research showed possible application as:

- slow release fertilizer exploiting the pyrolysis oil's high content in carbonyl groups through enrichment with nitrogen [54, 194, 199],

- preservative e.g. of wood [162] as some terpenoid and phenolic compounds present in pyrolysis oil act as insecticides and fungicides [194],

- insecticide to control insect pests in agriculture because of its toxicity (especially lignin oil) [200] and

- capturing agent of SO$_x$ in coal combustors (use as calcium salts and phenates from reaction of carboxylic acids and phenols with lime) [194].

Single chemicals of interest from pyrolysis oil are anhydrosugars like levoglucosan (for manufacturing of pharmaceuticals, surfactants, biodegradable polymers) [54, 162, 194], hydroxyacetaldehyde [54], or levoglucosenone, which has potential use in the synthesis of flavor compounds and antibiotics [194]. Only a few commercial niche applications exist. The most important are flavorings, essences and browning agents for food industry, also called liquid smoke. [53, 54, 194] The most important application barriers are low market value of chemicals at high production, separation, and purification costs, most for chemicals for which no established market exists [53, 54], the high separation and purification requirements arise primarily from the biomass and liquid product inhomogeneity [20]. Furthermore, nested interests (e.g. use as fuel vs. chemical production), scale of logistics (e.g. $10\,MW_{el}$ with $35\,\%$ electrical efficiency would need $40\,000\,t/a$ biomass feedstock, i.e. $40\,km^2$ arable land), high risk for new processes as well as public perception (not-in-my-backyard syndrome) impediment commercial use [53, 54].

2.2.4.2 Pyrolysis gas

For fast pyrolysis at moderate temperature and low vapor residence time, yields in the order of 10 wt.-% and for slow pyrolysis at low temperature and high vapor residence time around 35 wt.-% are typical [36]. The biomass feedstock composition influences the gas yield. Feedstocks rich in lignin pyrolyze to form less gas. Qu et al. [102] investigated the gas evolution of cellulose, xylan, and lignin pyrolysis in a tubular reactor. In the temperature range between 350 and 550 °C the yield was highest for xylan, increasing linearly from about 17 to 28 wt.-%. Above 550 °C the yield from cellulose was highest, reaching a maximum of 44 wt.-% at 650 °C. For lignin, on the other hand, the yield increased only from 8 to 22 wt.-% in the overall temperature range of 350 to 650 °C. At considerably higher temperatures (e.g. 1000 to 1400 °C [201]) the fuel is mainly gasified.

Pyrolysis gas contains CO, CO_2, H_2 and light hydrocarbons such as CH_4, propane, propylene, butane, butenes, C_5, ethane, etc. [102, 114, 123, 162]. Furthermore, it may contain SO_2, especially when produced from Kraft lignin (originating from pulping with a NaOH-Na_2S mixture) [42]. The higher heating value of pyrolysis gas is in the range of 15 to $20\,MJ/m^3$ [55], or 5 to 17 MJ/kg while the higher heating value of the gas from the three biomass components increases in the following order: cellulose < lignin < xylan with 1.3, 8.2 and 37.6 MJ/kg, respectively [193]. The gas can be used for (pyrolysis) process energy or as carbon feedstock (CO_2) for industrial applications [162].

2.2.4.3 Pyrolysis char

Figure 2.7 shows the degradation curves for the biomass components cellulose, hemicellulose (xylan), and lignin. When heated, xylan decomposes first, followed by cellulose. Lignin's mass loss rate has its maximum at the highest temperature, but it also reacts over the broadest temperature range. The curves obtained in this work (Fig. 2.7) are in good agreement with the data measured by Órfão et al. [202]. They investigated the thermal decomposition of cellulose, hemicellulose, and lignin with a heating rate of 5 K/min in the temperature range of 25 to 900 °C in nitrogen. Cellulose decomposition started at 225 °C with the main degradation peak at $\sim$ 370 °C, xylose, and lignin decomposition started at 160 °C and 110 °C, respectively. Lignin decomposition spreads over a wide temperature range up to 900 °C.

The char yield is influenced by process parameters, i.e. pyrolysis temperature, solids residence time, heating rate (all three having impact on carbonization degree), and feedstock composition like carbohydrate composition and mineral content. Generally, increasing pyrolysis temperature results in decreasing char yield leveling on a constant minimum value. When it comes to biomass composition, char yield increases from cellulose over hemicellulose (xylan) to lignin (Fig. 2.7 (a)) with 13.5, 24, and 41.7 wt.-% at 700 °C, respectively. Similar values can be found in literature [102, 202]. It should be noted that the lignin source and pulping conditions/process have an influence on the char yield [203, 204] (cf. 2.1.2). As the minerals contained in the feedstock are mainly retained in the char an increasing minerals content directly results in an increased char yield [102].

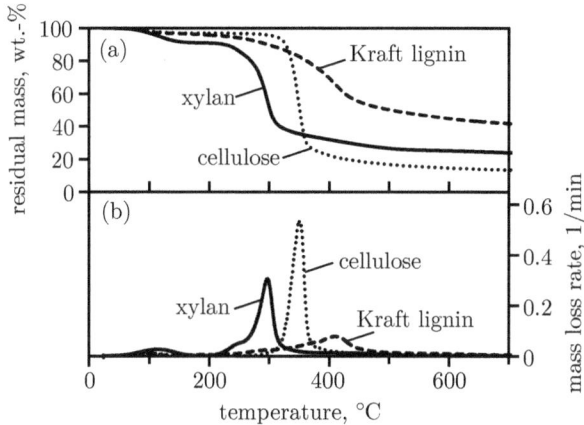

Figure 2.7: TG (a) and DTG (b) curves of cellulose, Kraft lignin, and xylan (hemicellulose model compound)
measured by *Mettler Toledo TGA/DSC 1* with N_2-flow of 40 ml/min, $\beta = 20$ K/min

Char contains (elemental) carbon together with hydrogen and oxygen, depending on its degree of carbonization. In addition, char also comprises the inorganics found in the pyrolyzed biomass (cf. Section 2.2.4.1). Usually, the content of single mineral species ranges from few 100 mg to several 10^4 mg/kg [189]. Sharma et al. [123] and Goyal et al. [162] reported char densities ranging from less than 300 to about 2000 kg/m^3. The measured densities (apparent and solid) depend on the degree of carbonization, and additionally on biomass type, pyrolysis conditions, and others [205–208]. The typical higher heating value of char (obtained from a large variety of biomass) is in the range of 24 to 33 MJ/kg, depending mainly on biomass composition [193, 209]. The pyrolysis temperature seems to have little effect [128]. Possible char applications are solid fuel in boilers, production of activated carbon, carbon fibers, carbon black, carbon-nano-tubes, soil amendment, nutrient adsorber, and direct supply of (pyrolysis) process energy [20, 162, 195].

2.3 Process modeling

2.3.1 Reaction pathway and kinetics

The reaction mechanism of biomass pyrolysis consists of numerous complex reactions with hundreds or thousands of intermediates and products. Therefore, it is currently not feasible to model the pyrolysis process in its entity, let alone all chemical reactions. Hence, reaction modeling on the basis of lumped or pseudo-components (char, gas, oil) is the widespread practice. [129, 210–212]. To consider biomass composition, the reaction mechanism and its kinetic parameters are either only valid for a specific feedstock or can include the dependence on the composition (cellulose, hemicellulose, and lignin) via parallel reaction schemes [212]. The pseudo-component reaction schemes can be summarized by a sequence consisting of an optional preliminary activation step (initial depolymerization), parallel primary reactions, and secondary cracking of mainly oil vapors (cf. Figure 2.8). Depending on the model complexity, some reaction steps may be omitted. The kinetic parameters are derived from experimental data and vary with the experimental conditions, i.e. temperature and heating rate. Consequently, the kinetic parameters have to be carefully selected for modeling a specific biomass pyrolysis reactor case. The pyrolysis reactions are typically assumed as first order and their temperature dependence described by the Arrhenius approach.

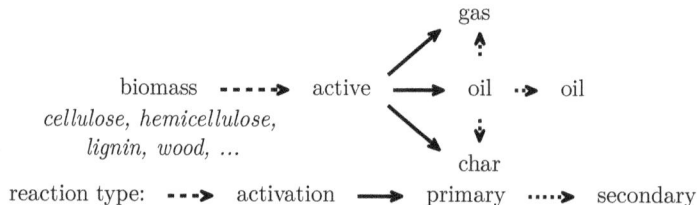

Figure 2.8: General reaction scheme with pseudo-components for biomass pyrolysis (combined and generalized from [91, 213–216])

An overview of the existing biomass reaction schemes is given in reviews [129, 210–212]. Examples of biomass reaction schemes can be found for various biomasses, i.e. spruce, eucalyptus, poplar, oak, beech, pine, corn, sunflower, and straw [91, 213, 216]. Unfortunately, for different biomasses, all kinetic parameters must be measured separately. Thus, interest in pyrolysis of biomass components increased, aiming to predict the pyrolysis behavior of biomass according to its composition, mainly cellulose, hemicellulose, and lignin but also extractives and ash, with cellulose being investigated most intensely. An overview of cellulose reaction mechanisms is given by Serbanescu [217]. Today, the Broido-Shafizadeh-model [218] is generally accepted and used for simulations [210, 215]. It was developed by Broido et al. [219] for cellulose and simplified by Bradbury, Allan G. W. et al. [218]. Based on this scheme, a model for biomass pyrolysis with superimposition of cellulose, hemicellulose, and lignin including secondary reactions from oil to gas was developed [220]. Another three-component model of superimposed reactions was developed by Miller and Bellan [221], which is predicated on another cellulose pyrolysis model [222] and extrapolated for the components hemicellulose and lignin. The amount of char formed from oil is small so that it is possible to neglect this reaction (cf. Figure 2.8) [213, 214, 223].

Due to possible interactions between the components, it is necessary to investigate the validity of the three-component-model. Órfão et al. [202] carried out TG analysis of the components cellulose, hemicellulose, and lignin as well as pine and eucalyptus wood. From the obtained data, a kinetic model with three independent reactions for the three components was derived and it was shown that the three-component-model can reasonably predict the volatile yields of the two investigated wood species. However, for pine bark the high content of extractives causes deviations. Furthermore, Peters [224] showed that the three-component-model can predict devolatilization of pistachio shells. Qu et al. [102] reported that in the case of rice straw, corn stalk, and peanut vine the additive three-component-model can predict the total yields of char, oil and permanent gas, as confirmed by Miller and Bellan [221] and Couhert et al. [225]. However, while the overall yields of char, gas, and oil can be predicted from the three biomass components, gas and oil composition cannot. The complex reactions underlying pyrolysis of cellulose, hemicellulose, and lignin lead to interactions between these three components and thus influence gas and oil composition. TG-FTIR analysis of samples with varying content of cellulose, hemicellulose, and lignin by Wang et al. [101] revealed for oil that formation of 2–furfural and acetic acid is enhanced by the presence of cellulose and lignin and the amount of 2,6-dimethoxy-phenol is increased by the combined influence of cellulose and hemicellulose. Furthermore, the interaction between cellulose and hemicellulose promotes the formation of 2,5-diethoxytetrahydrofuran, and inhibits the formation of altrose and levoglucosan. The presence of cellulose enhances the formation of hemicellulose-derived acetic acid and 2-furfural, whereas lignin is inhibiting. The gas component yields cannot be predicted from an additivity law as the components interact during pyrolysis. It was deducted by Couhert et al. [225] that interactions occur in the gas phase and probably also inside the solid reactant since the variation of cellulose, hemicellulose, and lignin content has an impact on the CO_2 yield. Raveendran et al. [226] suggest that the ash content in the biomass should be considered by the additivity law.

Models with increasing complexity exist, which include various chemical product species. An example is a pyrolysis scheme derived for Kraft lignin, which considers e.g. gaseous species like CO_2, multi-ring phenolics like phenanthrene, fluorene, bibenzyl in addition to light liquids, single ring phenolics, and carbonaceous residue [227, 228]. Another model considers the evolution of gaseous species like H_2, CO, CO_2, various C_mH_n, and vapor phase species like phenol, glyoxal or 5-hydroxy-methyl-furfural for beech wood or softwood pyrolysis [229]. However, these models are mainly derived for a single biomass (component) and for specific operating conditions and are difficult to generalize due to interactions between biomass components and the influence of pretreatment conditions.

2.3.2 Pyrolysis process modeling

A large number of pyrolysis process models with varying complexity have been developed. In most cases, the reactor models are based on fluid dynamics and single biomass particle pyrolysis. Single particle pyrolysis models have been developed as 1-D and 2-D models considering single wood particle or biomass component pyrolysis with reaction kinetics, particle, and product mass as well as heat balance [214, 216, 221, 230]. The model of Haseli et al. [214] also includes the temperature dependency of the heat of reaction. These models are either designed for micro-particle processes, thus neglecting intra-particle heat and mass transfer implying kinetic control or can additionally comprehend the influence of macro-particles for which internal diffusion and heat conduction resistance leads to concentration and temperature gradients inside the particle. Therefore, the product distribution may be influenced considerably by particle size [221, 230]. For evaluation of the governing regime, i.e. kinetic control or heat and mass transfer control, different criteria exist. Important in this case is the Biot number $Bi = h_T \cdot d_p / \lambda_T$, which correlates the external heat transfer (convection) to the internal heat transfer (conduction) of a body (a particle with diameter d_p). If $Bi < 0.1$ conduction is considerably greater than heat transfer to the particle. Thus, the particle temperature is almost uniform and the reaction kinetically controlled [144, 209]. For $Bi > 0.1$ a temperature gradient develops inside the particle body. The pyrolysis process is then governed by heat and mass transfer inside the particle. For typical biomass pyrolysis conditions in fluidized beds, this criterion can be reduced to the particle size. Scott et al. [231] pyrolyzed Avicel cellulose and maple sawdust in a fluidized bed and a transport reactor. They found that if the particle weight loss is smaller than 10 % before the particle reaches a core temperature of 450 °C, the reactor temperature will be the only variable determining the yields of char, oil, and gases for a given feed material and residence time. Kinetic consideration showed that particles with a diameter below 2 mm show this behavior. The boundary between kinetic and heat transfer control depends mainly on pyrolysis temperature and particle size [232] as well as heating rate [233]. For cellulose particles with a particle diameter below 200 μm and a pyrolysis temperature below 500 °C, the pyrolysis reaction is kinetically limited [232]. It was estimated that for lignocellulosic biomass with a particle size <1 mm heat and mass transfer phenomena are drastically decreased and kinetic control prevails [220]. Finally, fast pyrolysis of cellulosic particles is largely controlled by the rate of external heat supply, with internal heat transfer control becoming important for $T \geq 700\,\mathrm{K}$ and particles thicker than 3000 μm [131].

Models for fluidized bed reactor pyrolysis can predict the product distribution depending on fluidization and process parameters such as fluidizing velocity and pyrolysis temperature. The simplest model is a residence time model with no distinct fluid dynamics. Using such a model, it was shown that the influence of residence time and temperature can reasonably be described for poplar wood [213]. More complex models have additionally implemented the fluid dynamics of bubbling fluidized beds and the overall mass and heat transfer equations. Based on the kinetic scheme for the three main biomass components cellulose, hemicellulose, and lignin [221] and the fluid dynamics of a fluidized bed reactor [234] such a model was developed [235, 235]. However, the model is neither validated for application in circulating fluidized bed simulation nor for lignin pyrolysis. Also, CFD multi-fluid models with multi-phase fluid dynamics based on kinetics by Miller and Bellan [221] show reasonable results for fluidized bed pyrolysis. Modeling and experimental investigation has been reported by several researchers at University of Iowa [236–238], with a modeling time of about 200 s. Also for circulating fluidized beds, reactor models have been derived. A simple residence time model with no distinct fluid dynamics for different biomasses shows again that the influence of residence time and temperature can be calculated [91]. A 1-D steady-state CFB reactor model with kinetics for cellulose, hemicellulose, and lignin with mass and heat balancing equations shows that the fluid dynamics and yields for biomass with varying composition can be predicted and this model is applicable for use in flowsheeting [239]. In summary, models calculating the main fluid dynamics and product yields for either bubbling/stationary fluidized beds or circulating fluidized beds have been developed. However, distinct flowsheet modeling of lignin CFB pyrolysis is not known to the author of this work.

2.3.3 Flowsheet modeling of pyrolysis refineries

Increasing interest in sustainable lignocellulosic biomass use has brought about biorefinery concepts including pyrolysis processes, especially for lignin utilization. The simplest concept is solitary pyrolysis with product separation and collection – which was modeled e.g. by Qu et al. [102] in Aspen Plus®. A more complex example is the "BiomassPyrolysisRefinery"-concept proposed by Schwaiger et al. [240]: the pyrolysis oil is upgraded by deoxygenation and the char hydrogenated to liquid fuel and both refined together. Furthermore, biomass fractionation for cellulose and hemicellulose derivative production and successive lignin residue pyrolysis is proposed in literature [20, 29]. A pyrolysis refinery concept containing numerous oil upgrading and refining steps was simulated with Aspen Plus® [241, 242]. These models have in common that the pyrolysis process is depicted by a simple black box model (RYield block in Aspen Plus®) [102, 241, 242] which does not consider reactor specific fluid dynamics or heat and mass transfer phenomena. Thus, pyrolysis (biorefinery) flowsheet modeling including a comprehensive fluidized bed reactor model (both BFB and CFB) has not yet been covered in depth in literature.

3 Scope of work

Lignin pyrolysis, especially in fluidized beds, has proven to be difficult because of bed agglomeration or even clogging of feeding systems. While most research groups described these problems, others could finally pyrolyze lignin, e.g. by mechanical agitation of the fluidized bed. But also the lignin composition seems to have an influence on the pyrolysis behavior. To correlate possible feedstock influences a softwood Kraft lignin with a high purity and a hydrolysis lignin obtained from hydrothermal and subsequent enzymatic hydrolysis of wheat straw with a low purity are characterized and pyrolyzed. To prevent the reported agglomeration, defluidization and feeding problems a circulating fluidized bed (CFB) system with cooled pneumatic feeding was designed, erected and operated. As an in-depth investigation of CFB lignin pyrolysis is not known to the author, the composition of pyrolysis char, oil and gas are measured and their composition and yield determined.

To shed some light on the agglomeration and clogging issues, the two used feedstock are characterized by Flash-DSC at heating rates comparable to heating rates in fluidized bed applications. Moreover, the obtained characteristics are compared to reference materials by the same method: cellulose, xylan and straw lignins attained by organosolv (high purity) and soda (intermediate purity) pulping. Furthermore, the morphology of the solid samples from the CFB (a mixture of quartz sand and char) is analyzed to understand the difference in pyrolysis behavior and mechanisms in char particle formation.

The pyrolysis performance is determined by measurement of the overall yields of the pseudo products gas, oil, and char as well as their composition. The obtained yields are correlated to the main process parameter, which is the pyrolysis temperature. The circulating fluidized bed system works semi-continuously, i.e. gas and oil vapors are continuously removed from the hot reaction zone by the carrier gas, while the char remains mostly in the CFB system. With the char also the feedstock ash accumulates in the CFB system. Due to this accumulation and the catalytic activity of some minerals, an influence on product distribution can be expected. Therefore, samples at different experiment duration and times have been taken for all products and were subsequently analyzed.

To reach a broader understanding of the pyrolysis process the experimental work was accompanied by process simulation. For flowsheet simulation of integrated refineries or pulping plants including a CFB pyrolysis process for lignin utilization, a CFB lignin pyrolysis unit model is necessary. Problems of previous models are that often only limited feedstock types and/or reactor conditions can be accounted for. But in an integrated process, which the flowsheet simulation is supposed to represent, the input parameter may vary greatly. For example, the feedstock composition and lignin purity obtained by pulping might change due to different pulping parameters. Furthermore, it might be desired to operate the pyrolysis process in circulating or stationary regime. Thus, it is the aim to develop a model, which is predicting pyrolysis process yields for a broad pyrolysis feedstock composition (by means of cellulose, hemicellulose and lignin content) and works for both circulating and bubbling fluidized beds considering the fluidized bed fluid dynamics. Furthermore, this broad applicability to various biomasses and fluidized bed systems has the advantage, that the model can be validated by experimental literature data not readily available for lignin.

Sensitivity analyses on discretization, fluidization velocity, and biomass composition show the functionality of the model. The model is not only validated by the data obtained in this work but also literature data for pyrolysis of maple wood, poplar wood and wheat straw in a bubbling fluidized bed as well as pine wood in a circulating fluidized bed. Contrary to the fact that the oil composition obtained from cellulose, hemicellulose, and lignin pyrolysis is fundamentally different, the important secondary reactions for conversion of oil to gas are mostly assumed to be equal. Therefore, the kinetic data sets for lignin are compared to the experimental data obtained in this work. Subsequently, a proposal on model improvement is given and the model goodness validated by comparison with the mentioned own and literature data.

For proof of applicability to flowsheet simulation as well as energetic process evaluation the simulation of a process integrating char and pyrolysis gas combustion is attempted. Therefore, a flowsheet containing the pyrolysis reactor, a fluidized bed char combustor and short-cut models for gas-solid separation, oil condensation, permanent gas combustion as well as heat exchangers is set up with mass and enthalpy balances. The pyrolysis process efficiency is evaluated by the oil energy recovery rate and the integrated pyrolysis-combustion-process is rated by a surplus-to-deficit ratio.

4 Experimental

4.1 Pyrolysis plant

The pyrolysis plant consists of three main components: circulating fluidized bed (CFB) reactor, liquid (oil) product separation and after burning chamber (ABC) (cf. Figure 4.1).

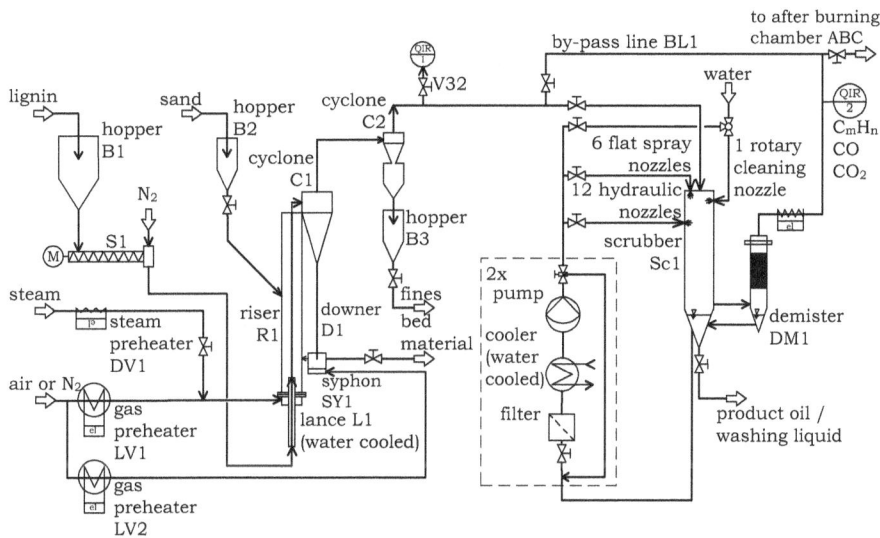

Figure 4.1: Scheme of pyrolysis plant (after burning chamber (ABC) not depicted)

Pyrolysis system

The heart of the pyrolysis plant is the circulating fluidized bed reactor, with its scheme depicted in Figure 4.1 and in detail in Figure 4.2. The fluidizing gas is preheated electrically by preheater LV1 to 350 °C and introduced via a windbox and a bubble cap gas distributor into riser R1 (80 mm diameter). The distributor has three triangularly positioned bubble caps. As fluidizing gas air, nitrogen, steam or a mixture of nitrogen and steam can be used. Air and nitrogen are supplied at 5 bar by a compressed air supply and bottled pressurized gas, respectively. Steam is supplied by a controllable steam generator (max. 9 kW) at 2.2 bar and 150 °C, an insulated pipe, and a superheater DV1. The bed material is fluidized in fast fluidization regime so that is entrained at the top of the 1700 mm high riser. The particle laden gas enters the primary cyclone C1 and most of the particles are separated from the gas. The particles are returned to the riser R1 via a standpipe D1 (53.1 mm diameter) and a syphon SY1 (100 mm diameter). The syphon is fluidized with air or nitrogen at about 0.15 m/s superficial gas velocity. Downstream from cyclone C1 fine particles are separated from the gas stream by a secondary cyclone C2. The cyclone dimensions are given in Figure 4.3. The whole reactor system is made of stainless steel (1.4841) and heated electrically. Refractory lining in not used to ensure short heating and cooling time and thus flexible operation.

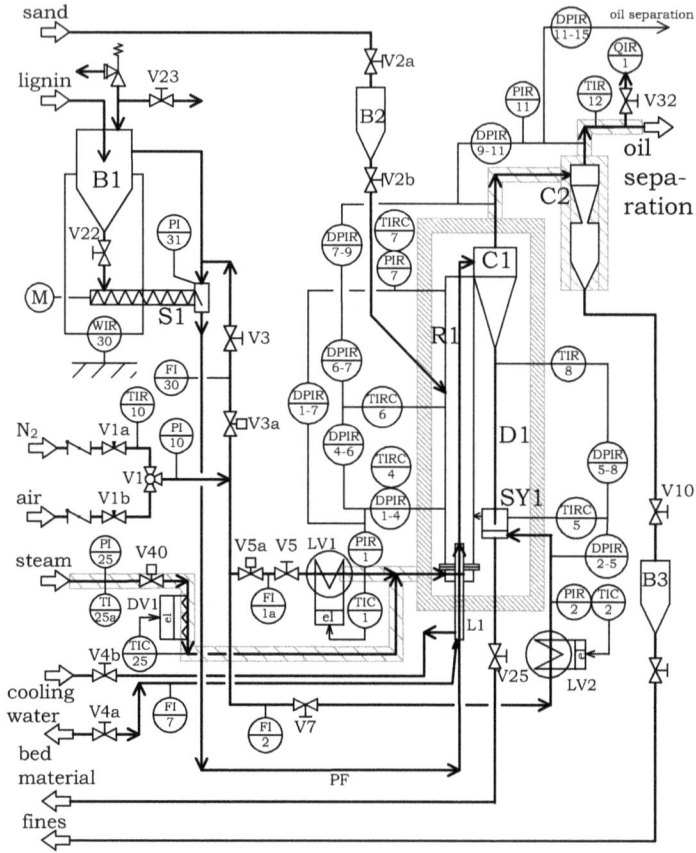

Figure 4.2: Scheme of pyrolysis reactor

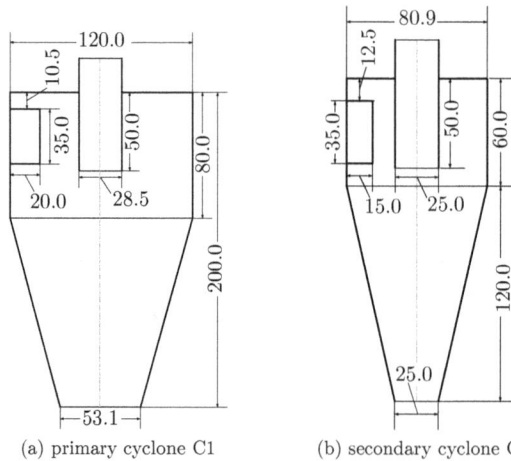

(a) primary cyclone C1 (b) secondary cyclone C2

Figure 4.3: Sketch of cyclones (all dimensions in mm)

Oil separation system

Gas and oil vapor exiting the reactor system via C2 are cooled rapidly in the oil separation part of the plant, depicted in Figure 4.1 and in detail in Figure 4.4. By cooling the gas and oil vapor mixture rapidly secondary reactions are quenched and the oil is condensed. Quenching is achieved by a scrubber Sc1, which is operated with water. Water is recirculated from the scrubber sump by two lines consisting of a liquid filter with two chambers (F1 and F2), a plate type heat exchanger (W1 and W2) and a membrane pump (P1 and P2) each. The pump of line one is pumping the water through two rings of six hydraulic nozzles each into the scrubber gas volume. Further, a rotary cleaning nozzle is operated to clean the scrubber walls from tars. The second line is supplying water to six flat spray nozzles, which produce a liquid film on the vertical scrubber walls. Additionally, the third ring of six hydraulic nozzles is connected to line two. The hydraulic nozzles mainly achieve the cooling duty and are operated at 10 bar. The rotary cleaning and the flat spray nozzles are operated at 2.5 bar and at 2 bar, respectively. The gas exiting the scrubber Sc1 passes through a demister and is heated to a temperature of more than 150 °C to prevent condensation in the pipe to the ABC and consequently blocking. The scrubber Sc1 can be by-passed by BL1.

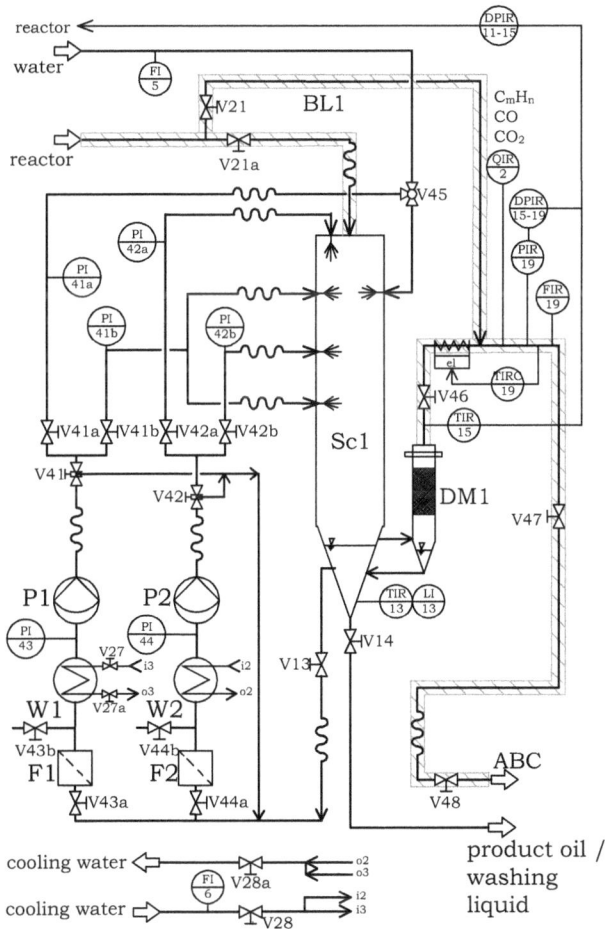

Figure 4.4: Scheme of oil separation

After burning chamber

From the oil separation system, the gas is introduced into a circulating fluidized bed combustor CFBC functioning as an after burning chamber ABC. The gas is fed at 500 mm above the gas distributor into the CFBC riser. The CFBC which is operated with air, has a height of 15 m, and a diameter of 100 mm. The combustion takes place at 850 °C. Downstream the after burning chamber the flue gas is cooled and mixed with fresh air before entering the induced draft fan. A detailed description of the after burning chamber with subsequent cooling and fan is given in [243, 244].

Lignin feeding system

Lignin is fed axially into the riser R1 from the bottom through a water cooled feeding lance L1 (cf. Figure 4.5). The lance is positioned coaxially between the bubble caps and its tip is positioned 140 mm above gas distributor plate. Water cooling is necessary to prevent softening of the lignin and consequently blocking of the feeding line. The lignin is conveyed pneumatically by nitrogen through the lance (6 mm inner diameter) at 35 m/s. The lignin mass flow is controlled by a screw feeder with adjustable rotor speed. The average mass flow is monitored by weighing the whole system of lignin bunker and screw feeder.

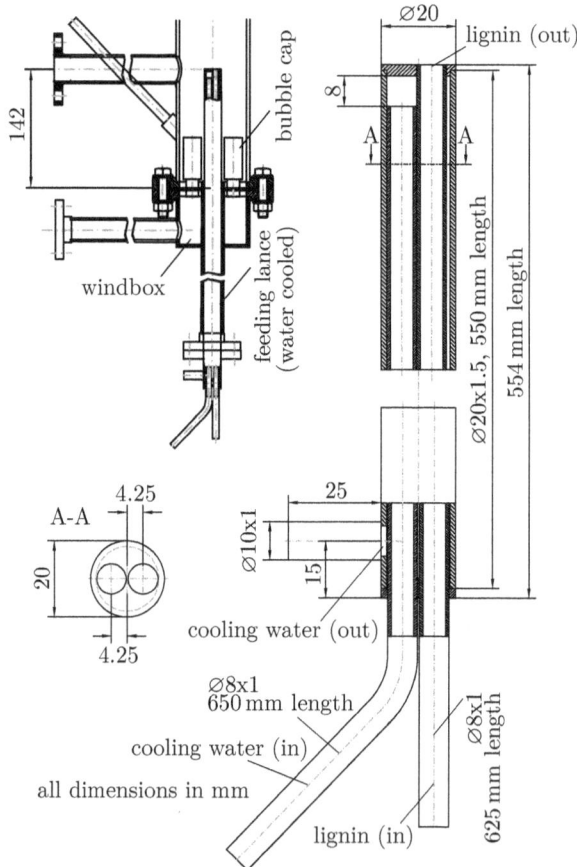

Figure 4.5: Lignin feeding lance: detail drawing and installation at riser bottom (top left)

Online measurement system

For the operation of the pyrolysis plant, online measurements of pressure drops, absolute pressure, temperatures and gas concentrations are carried out. The acquired signals are stored and displayed by the system design software LabView. The pressures and pressure drops are obtained by 26 differential pressure transducers. These transducers have different measurement ranges: 0 to 1035 mbar and 0 to 345 mbar for absolute pressures (measured relative to ambient pressure) and 0 to 345 mbar and 0 to 69 mbar for relative pressures, i.e. pressure drops. The measured pressure drops in the CFB system indicate the solids recirculation and the pressure drop over the scrubber Sc1 acts as an indicator of blockage. The pressure signals are stored at 1 Hz. Temperatures are measured with type K thermocouples. In total 15 temperature sensors are installed and the signals are processed by a 3 times 5 channel multiplexer (operated at 3 Hz) and an analog-to-digital converter. Also the temperatures are stored and displayed with a frequency of 1 Hz. For online measurement of the permanent gas composition, a side stream is taken between the demister DM1 and the after burning chamber ABC at QIR2 (cf. Figure 4.4). The measurement system is shown in Figure 4.6. The side stream is taken using open tubular-probes. In an electrostatic precipitator, which is operated at 3 kV, the gas is cleaned from tar. A heated filter (200 °C) separates fines before the gas is dewatered in a cooler. A flame ionization detector (FID) of type RS55-T (Ratfisch Analysensysteme GmbH) is used to analyze the C_mH_n concentration of the gas. The FID-unit has a built-in gas filter, is fueled by hydrogen and the by-pass stream of the unit is used as the feed for the subsequent analyzers. The measurement range of the FID changes dynamically within five ranges, with a total range of 0 to 100 000 ppm. Downstream the FID-unit sequentially coupled (nondispersive infrared NDIR) GME.42-analyzers (TAD Gesellschaft für Elektronik-Systemtechnik mbH) for CO_2 and CO are used. The analyzers have a range of 0 to 20 vol.-% and 0 to 10 vol.-%, respectively. The accuracy of all three analyzers is 2 % of the upper range value. The volume flows for the analyzers are controlled by rotameters. The off-gas of the measuring system is sent back to the plant off-gas.

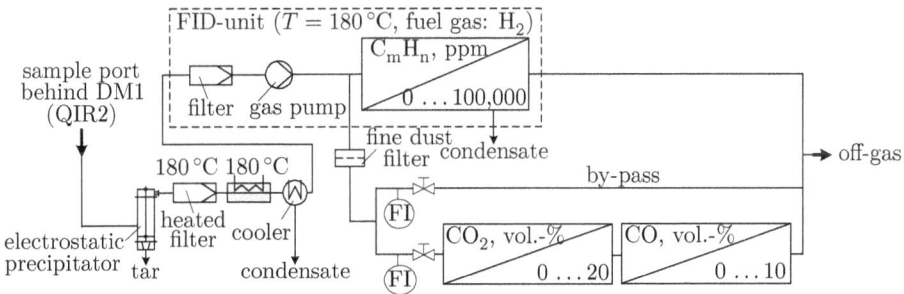

Figure 4.6: Measuring system for permanent gas concentration at QIR2

Operating procedure

Start-up

Prior to the experiment, the lignin is dried at 60 °C in a drying cabinet over night. 5 to 10 kg lignin are filled into hopper B1 and the weighing is started. The scrubber system is filled with water and deaerated by V43b and V44b. Cooling water for the scrubber system and the feeding lance is switched on. 5 kg quartz sand is filled into the CFB system. During heating up the scrubber is operated in bypass mode. The air pressure reduction valve is adjusted to 5 bar and the feeding air is turned on to prevent clogging of the feeding lance. Then, the riser fluidization air and syphon are adjusted to 4 to 5 m/s and 0.1 m/s, respectively. The solids recirculation is monitored such that the total pressure drop over distributor and bed DPIR 1-7 adjust itself between 30 and 120 mbar and the pressure drop in the upper part of the riser DPIR 6-7 between 3 and 15 mbar. The oven, secondary cyclone, and pipe heating are adjusted to the desired set point. Subsequently, the gas preheaters LV1 and LV2 are turned on (max. 500 °C). When the set point temperature in the riser are reached, the steam generator and heating of steam pipe are turned on and the air flow is reduced in such a way that the superficial gas velocity in the riser is maintained at 4 to 5 m/s. Thereafter, the nitrogen bottle valves are adjusted to 5 bar and the three-way valve V1 switched to N_2. The lignin hopper is flushed with nitrogen by opening V23 using counterflow through the screw. When the oxygen has been removed from the plant the scrubber is set into operation (the bypass is closed and scrubber connection opened). To start pyrolysis operation the screw feeder is turned on and adjusted to the desired set point. Afterward, the voltage of the electrostatic precipitator might need adjustment.

Stable operation

During operation, the main parameters of feeding system, CFB system, oil separation system, and after burning chamber are controlled. Important parameters are: feeding rate, pressure at screw feeder outlet (clogging of lance), CFB temperatures, and pressure drops (regular solids circulation and heating), maximum pressure in riser and pressure drop over the scrubber system (clogging in oil separation system) as well as pressures in scrubber recirculation lines (proper operation of liquid injection).

Shutdown

For shutdown, the lignin feeding is turned off and it is waited until pyrolysis is complete (no pyrolysis gases measured by online analyzers). All heating devices and the steam generator are turned off. Subsequently, the N_2 supply to the syphon is stopped to stop solids recirculation and accumulate the bed material in the syphon and standpipe. When the pressure drop over distributor and bed DPIR 1-7 dropped to the distributor pressure drop and the pressure drop in the upper part of the riser DPIR 6-7 to 0 mbar the N_2 flow to the riser is reduced to 2 m^3 (standard conditions). By the time the CFB system temperature has come down to 300 °C the nitrogen supply is closed.

4.1.1 Pyrolysis product sampling

Pyrolysis oil sampling

Consecutive side stream samples are taken at different pyrolysis times during the experiments. The sketch of the sampling system is shown in Figure 4.7. At V32a between secondary cyclone C2 and Scrubber Sc1, an open tubular-probe is inserted orthogonally into the gas stream and centrally arranged in the pipe (cf. Figure 4.8). The sample train consists of the open tubular-probe (including V32), one 500 ml and five 250 ml impinger bottles (cf. Figure 4.9a). The first five bottles are filled to approximately equal level with in total 1 l isopropanol. These bottles are cooled to −19 °C in an ice bath. Bottles two to four are equipped with glass frits for good gas distribution. The insert of the first bottle has a plain stem to prevent clogging. The pyrolysis oil is condensed in the first four bottles. The fifth bottle serves as blank test to prove that the oil is completely condensed in the first four bottles. The sixth bottle is filled with cotton wool for adsorption and demisting (cf. Figure 4.9b). Downstream the sample train temperature, pressure, volume flow, and oxygen content are measured for balancing purpose. Before sampling the sample train is inertized by N_2 purge gas (connection of purge gas to silicone connection). The oxygen level is monitored by the Oxynos 100 O_2-analyzer (Emerson Process Management GmbH & Co. OHG). The measuring range of the analyzer was adjusted to 0 to 23 vol.-% and an accuracy of $\leq 1\%$ of limit of detection. When the oxygen concentration has reached zero, the sample train is attached at the silicone connection to V32 and the sampling is started. Up to 20 l (standard conditions) sampling gas is sucked through the sample train with a volume flow of 20 l/min. Although the impinger bottles are sealed with grease, the system cannot be sealed completely from the environment at any time. Therefore, the oxygen concentration is continuously measured also during sampling. After sampling is complete, V32 is closed and the sampling time is manually recorded. Then, the open tubular-probe (including V32) is exchanged, as an oil fraction does already condense on the surface of the tube and valve. At that moment, the sampling routine can be repeated.

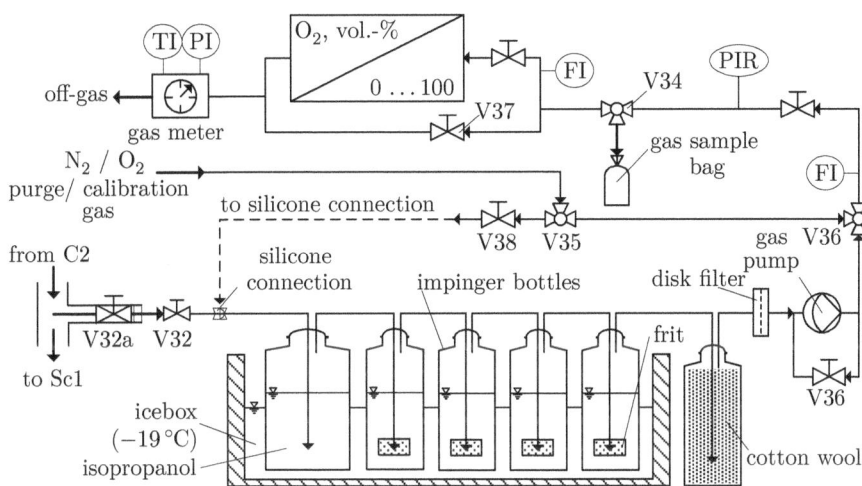

Figure 4.7: Side stream sampling system for pyrolysis oil (QIR1)

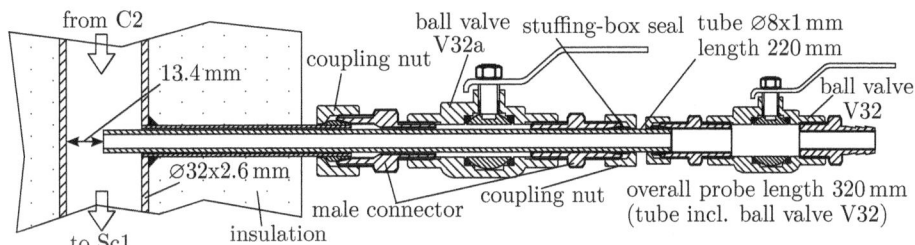

Figure 4.8: Sampling probe V32 (not drawn to scale)

(a) Sample train B of Exp. V89

(b) Sampling during Exp. V89

Figure 4.9: Pictures of side stream sampling

Gas sample bags

As the online gas measurement is carried out for CO, CO_2 and C_mH_n no information about H_2 and hydrocarbon composition is measured online. To gain this additional information, at the end of each oil sampling a gas bag is attached to V34 (cf. Figure 4.7). It is filled with up to 10 l (standard conditions) and the content is analyzed by GC.

Char sampling

After plant shutdown, when ambient temperature is reached, the bed material (BM) is discharged from the syphon. The material is weighed (for balancing purpose) and a sample taken. Additionally, the material from the hopper B3, below secondary cyclone C2, is weighed and a sample taken. This latter material is named secondary cyclone material and abbreviated C2 in this work. Both samples are a mixture of initial bed material (quartz sand) and pyrolysis char.

4.2 Analysis methods for characterization of used material and pyrolysis products

4.2.1 Sample preparation

The pyrolysis oil samples are dissolved in isopropanol and for analysis have to be separated from the solvent. The bottles 1 to 4 are collected together with the product which condensed in the tubular-probe and valve V32. Rinsing of the tubular-probe and valve V32 is done with isopropanol and methanol. In parallel, the fifth bottle is prepared for analysis analogously. That way it can be guaranteed that all dissolvable oil components are collected in the first four bottles. The water contained in the sample has to be separated before analysis. For drying of the organic solution Na_2SO_4, is added in excess. The preparation for analysis follows different pathways. These paths are shown in Figure 4.10 and the corresponding procedures and analyses were carried out at the Thünen Institute of Wood Research, Hamburg, Germany.

- Path 1: One fourth of the sample is concentrated in a Vigreux-distillation column at 82 °C (boiling point of isopropanol). The concentrated sample is transferred to a volumetric flask and filled with isopropanol to 20 ml. Then 10 ml of the sample are dried completely in a rotary evaporator under vacuum and 40 °C and the remaining tar2 is weighed. A fraction of the tar is then dissolved in dimethyl sulfoxide (DMSO) for SEC analysis.

- Path 2a: For GC analysis, 1 ml from the sample vial (path 1) is taken.

- Path 2b: As it is difficult to quantitatively determine catechols directly by GC, the catechols have to be derivatized for analysis. Therefore, another fourth of the sample is dried completely in a rotary evaporator, the remaining tar1 weighed and redissolved in pyridine. For derivatization BSTFA+TMCS 99:1** is added in excess to the solution. The silylation reaction causes the exchange of hydroxy groups by trimethylsilyl groups. The silylated catechols are easier to evaporate in the GC system and can be qualitatively and quantitatively determined. Tar1 is used also for ultimate analysis of the pyrolysis oil.

Furthermore, half of the sample is retained for additional or repeated analysis.

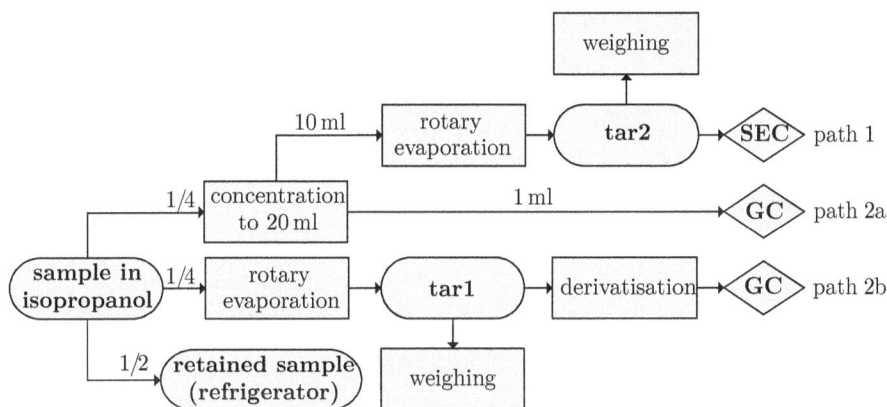

Figure 4.10: Sample preparation for in isopropanol dissolved pyrolysis oil from side stream sample train

4.2.2 Analysis methods

For evaluation of experiments as well as analysis of feedstock and products, various methods have been applied in this work. A short overview on analytic sample measurement methods is given in Table 4.1. The analytic methods are further described below.

Gas chromatography (GC)

*Permanent gas**
The gas in the sample bags (cf. Section 4.1.1, page 29) is analyzed by GC-TCD and GC-FID. A sample of 250 μl is injected into the GC-TCD Agilent 6890 GC (column flow rate: 43.9 ml/ min He). It is used for quantification of H_2, CO, CO_2, O_2, N_2, and CH_4. The GC-TCD is equipped with a Porapack Q column of 2 m, 1/8″, 80/100

**BSTFA (N,O-bis(trimethylsilyl)trifluoroacetamide) + TMCS (trimethylchlorosilane), Sigma Aldrich [245]

*Measured at Institute of Environmental Technology and Energy Economy, Hamburg University of Technology, Germany

Table 4.1: Applied analysis methods for used and produced materials

samp./meth.[†]:	PSD	PAM	SEM	UA	PA	TM	KFT	ρ	FD	CHL	GC	SEC
lignin	X		X	X	X	X		X	X	X		X
ref. biomass				X	X			X[‡]	X	X		X[*]
quartz sand	X		X					X				
bed material	X	X	X	X	X	X		X				
C2-material	X	X	X	X	X	X		X				
pyrol. oil				X	X	X	X				X	X
pyrol. gas											X	

[†]PSD: particle size distribution; PAM: physical adsorption measurement; SEM: scanning electron microscopy; UA: ultimate analysis; PA: proximate analysis; TM: trace metals by inductively coupled plasma optical emission spectrometry; KFT: Karl-Fischer titration; ρ: solid, apparent and/or bulk density; FD: Flash-DSC; CHL: carbohydrate composition and lignin content; GC: gas chromatography; SEC size exclusion chromatography; [‡]solid density only; [*]lignins only,

mesh and a molecular sieve 5A of 2 m, 1/8″, 60/80 mesh. It is operated in sample loop with an isotherm 50 °C SP1Temp program. Furthermore, a sample of 300 µl is injected into the GC-FID Chrompack CP 9000 (column flow rate: 8.6 ml/ min N_2). It is used for qualitative and quantitative analysis of additional hydrocarbons. The GC-FID is equipped with an Al_2O_3/S column of 50 m, ID 0.53 mm, d_f 15 µm. It is operated in splitless mode with the following SP1Temp program: 1 min holding at 76 °C, heating with 2 °C/ min to 100 °C, heating with 8 °C/ min to 148 °C, heating with 12 °C/ min to 200 °C and holding the final temperature for 32 min.

Pyrolysis oil[††]

For quantitative identification of single pyrolysis oil components, a known amount of the internal standard fluoranthene is added to the sample before injection (path 2a and 2b in Figure 4.10). As for all sample preparation pathways and steps the split rates and the component specific GC response factors are known, the single component mass in a side stream sample can be calculated. The single component content is measured by a GC-MS/FID Agilent 6890N with 5975B MS for liquid injection. It is operated with an Agilent VF-1701 60 m x 0.25 mm column (column flow rate: 2.0 ml/ min He). A 1.0 µl sample is injected with 15:1 split mode. The applied temperature program is: 4 min holding at 45 °C, heating with 3 °C/ min to 280 °C and 20 min holding at the final temperature of 280 °C (total: 102.33 min). For identification, the NIST2012 library and an in-house pure substance library are used. The MS signal is used for identification and the FID signal is used for quantification of a substance.

Size exclusion chromatography (SEC)[††]

Both the pyrolysis oil samples prepared as described (cf. path 1 in Figure 4.10) and the lignin samples were analyzed by the following procedure: The lignin is dissolved in DMSO with 2 % LiBr in a concentration of 1 to 2 mg/ml. Of the dissolved lignin/the pyrolysis oil a sample of 100 µl is injected into the SEC Agilent 1100. It is equipped with two Varian PolarGel-L columns, each with 300 mm length, a 7.5 mm ID and operated at 60 °C. The solvent flow rate is 0.8 ml/ min. The UV detector (254 nm) is used for signal measurement. Calibration is carried out with various polyethylene glycol standards.

[††]Measured at Thünen Institute of Wood Research, Hamburg, Germany

Particle size distribution (PSD)

Three different particle size measurement techniques are used for characterizing the particle size of the solid material:

- Beckman Coulter LS13320 laser diffraction particle size analyzer [246] with Fraunhofer evaluation method (sample dissolved in isopropanol (Kraft lignin, bed and secondary cyclone material))

- Camsizer XT dynamic image analyzer (Retsch GmbH) [247], X-Fall (gravity dispersion) with measuring range of 1 to 7000 µm (quartz sand, hydrolysis lignin, bed and secondary cyclone material)

- Sieve analysis, standard DIN66165 [248] (mesh sizes: 100, 300, 450, 710, 1600, 2000, and 4000 µm; for agglomerated Kraft lignin and Kraft lignin behind the screw feeder)

Scanning electron microscopy (SEM)[¶]

For determination of particle morphology, a scanning electron microscope of type Zeiss Supra 55 (Oxford Instruments) with SE2 and EDX detector is used. The acceleration voltage is adjusted between 3 to 4 kV and the measuring chamber pressure kept at 10^{-4} to 10^{-3} Pa. Prior to measurement, the lignin samples are sputtered with a 2.5 nm gold layer to prevent electrical charging of the sample. For analysis of single particle elemental surface composition of solid product particles, energy-dispersive X-ray spectroscopy (EDX) is applied.

Total lignin content and carbohydrate composition[††]

For determination of lignin content and carbohydrate composition, a procedure according to Willför et al. [249] and Lorenz et al. [250] is applied. Prior to borate-HPAEC for carbohydrate quantification, sample preparation is performed by 2-step hydrolysis. At first, a 200 mg sample is ground, followed by the first hydrolysis step where the sample is hydrolyzed at 30 °C with 2 ml sulfuric acid (72 % by volume). After one hour the hydrolysis reaction is stopped by addition of 6 ml distilled water. The sample is transferred to a volumetric flask and 50 ml distilled water are added. For the second hydrolysis step, the sample is autoclaved at 1.2 bar and 120 °C for 40 min. The hydrolyzate is filtered with G4 sinter glass frits (pore size 10 to 16 µm) and the lignin residue dried and weighed. Together with the mass of the acid soluble lignin (detected by UV-spectroscopy) the total lignin content is calculated. The liquid fraction is separated on an Omnifit column of 115 mm length and 5 mm diameter. The column is packed with a stationary phase, which is a potent anion exchange resin MCI Gel CA08F (Mitsubishi-Chemical) tempered at 65 °C. For the mobile phase, water based potassium tetraborate/boric acid-buffer is used in two concentrations: A (0.3 M, pH 8.6) and B (0.9 M, pH 9.5). The flow rate is adjusted to 0.7 ml/min. At the start of the elution program, the two buffers are mixed in a ratio of 90% A and 10% B and linearly changed within 35 min to 10% A and 90% B. After 8 min the ratio is linearly reversed to 90% A and 10% B within 7 min. After separation, the analytes are derivatized by Cu-bicinchoninate in a teflon coil at 105 °C and a flow rate of 0.35 ml/min. The teflon coil has a length of 30 m and diameter of 0.3 mm. Identification and data processing of the carbohydrates is executed via UV/VIS-detection operated at 560 nm and the Dionex Chromeleon software.

[¶]Measured at Central Division Electron Microscopy, Hamburg University of Technology, Germany
[††]Measured at Thünen Institute of Wood Research, Hamburg, Germany

Gas adsorption[‡‡]

The average pore size and surface area of the char and quartz sand are determined by gas adsorption measurement. Prior to the sorption experiments all samples are degassed for 20 h at 200 °C and vacuum conditions. Mesopore area and average pore size are determined by nitrogen sorption experiments (desorption branch) on a NOVA 3000e Surface Area Analyzer (Quantachrome Instruments, Boynton Beach, FL, USA). The analysis is carried out according to the IUPAC standard procedure [251]. The Brunauer-Emmett-Teller (BET) method is used for evaluation of the sorption data. Micropore area and average pore size are determined by CO_2 sorption experiments (desorption branch) on the same analyzer. The Dubinin-Radushkevich (DR) method [252] is used for evaluation of the micropore sorption data. Evaluation is done with the software Quantachrome NovaWin.

Elemental analysis

The water content of bed and secondary cyclone material is measured according to the German standard DIN51718 [253] at 106 °C. Due to decomposition of lignin above 66 °C the biomass samples are dried at 40 °C in a vacuum drying cabinet for determination of the water content.

The bed material, secondary cyclone material, and lignin samples are analyzed according to the German standard DIN51719 [254] for determination of the inert content. The sample is incinerated at 815 °C and the solid residue balanced. Deviating from standard fuels, bed and secondary cyclone material consist of a mixture of pyrolysis char and quartz sand. Thus this procedure gives the sum of the samples ash content originating from the char and the quartz sand content. Figure 4.11 illustrates the composition of the samples. Therefore, the inert content is differentiated in this work by inert, sand and ash content. The ash content of a bed or secondary cyclone material can only be approximated by calculation from the char yield and ash content of the lignin (cf. Section 4.3.3).

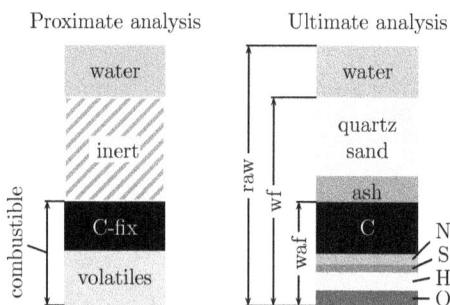

Figure 4.11: Composition of bed material

The elemental composition (C, H, N, and S) of solid samples, i.e. bed material, secondary cyclone material, and biomass, as well as pyrolysis oil is analyzed in a *elementar vario MACRO cube (Elementar Analysensysteme GmbH)*.[§] The oxygen content is calculated

---

[‡‡]Measured at Institute of Thermal and Separation Processes, Hamburg University of Technology, Germany

[§]Measured at Central Division Chemical Analytics, Hamburg University of Technology, Germany

From the elemental composition of biomass, char, and oil, the higher heating value HHV H_0 can be approximated. The correlation derived by Channiwala and Parikh [255] was not only fitted to coals but additionally to biomass, gaseous and liquid fuels. For H_0 it holds in MJ/kg:

$$H_0 = 34.91w_C + 117.83w_H + 10.05w_S - 10.34w_O - 1.51w_N - 2.11w_A \qquad . \qquad (4.1)$$

The ash composition of bed and secondary cyclone material, pyrolysis oil, and quartz sand are measured with ICP-OES (Inductively Coupled Plasma Optical Emission Spectrometry).[§] For analysis, the samples are dissolved in nitrohydrochloric acid. An Optima 7000 DV ICP-emission spectrometer (PerkinElmer, Inc., Boston, USA) with integrated software is used for analysis. The metal ions Fe, Zn, Mn, Al, Na, K, Ca, Mg, Cu, Li, Ni, P, and S are quantified.

Water content of pyrolysis oil[‡‡]

The water content of the pyrolysis oil can only be determined for experiments without $H_2O(g)$ as fluidizing medium. And even for experiments that are carried out with only nitrogen as fluidizing gas, the determination of the water that is produced in the pyrolysis reaction, is difficult as the water content of the feed material and the isopropanol in the sample train have to be considered. Also, the split rate for sampling (cf. Section 4.3.2) is very low, resulting in extremely low water content in the isopropanol sample. Nevertheless, for some samples (for experiments with only nitrogen as fluidizing medium) the water content is determined with a Karl-Fischer Coulometer C20/C30 (Mettler-Toledo Inc.) according to DIN51777-1 [256].

Density and porosity of solid material

The material, apparent as well as bulk densities are obtained by the calculation specification provided in DIN 66137-1 [257]. Measurement of the material density ρ_m is done for quartz sand, lignin, reference biomasses, bed and secondary cyclone material. The measurement is carried out with a MultiVolume Pycnometer 1305 (Micromeritics Instrument Corporation) working with He at 140 to 170 kPa and a sample size of 5 cm^3. The apparent density ρ_a for quartz sand, lignin, bed and secondary cyclone material is measured in a liquid pycnometer of known volume (50 ml). Therefore, Kraft lignin, bed and secondary cyclone material are dispersed in isopropanol, whereas hydrolysis lignin and quartz sand are dispersed in water. The density of the dispersing agent is taken from [258] at the measured temperature. The bulk density ρ_{bulk} of selected solid samples is measured following DIN51605 [259]. The char density ρ_{char} is calculated from the rule of mixture of quartz sand ρ_{QS} and bed ρ_{BM} or secondary cyclone densities ρ_{C2} as well as the mass fraction of char $w_{char, BM/C2}$ in the sample, respectively.

$$\rho_{char} = \frac{w_{char, BM/C2}}{\frac{1}{\rho_{BM/C2}} - \frac{1-w_{char, BM/C2}}{\rho_{QS}}} \qquad (4.2)$$

The porosity is calculated from material and apparent densities and the calculation specification provided in DIN 66137-1 [257].

[§]Central Division Chemical Analytics, Hamburg University of Technology, Germany
[‡‡]Measured at Institute of Thermal and Separation Processes, Hamburg University of Technology, Germany

Thermal analysis by Flash-DSC[‖]

Chip and instrument

For analysis of material behavior at high heating rates corresponding to pyrolysis conditions in a fluidized bed, a Flash DSC 1 (Mettler Toledo Inc.) is used. The Flash DSC 1 has a twin membrane calorimeter chip, based on MEMS (Micro-Electro-Mechanical Systems) sensor technology. The MultiSTAR UFS1 (24 x 24 x 0.6 mm^3) MEMS chip sensor (Figure 4.12(a)) is mounted on a ceramic base plate and is coated with aluminum to achieve a homogeneous temperature distribution. [260] Each membrane has a 0.5 mm diameter. One membrane holds the sample, whereas the other remains empty and serves as reference. The sensor has 16 thermocouples (8 polysilicon thermopile each on the sample and reference sides) [261]. Instead of using a sample crucible the sample particle is placed directly on the sample membrane. Thereby the required heating power is kept low and temperature inaccuracies at high heating rates are minimized. [260] Heating and cooling are controlled by dynamic power compensation [260], making measurement at a temperature range (ambient temperature + 5 K) to 500 °C with typical cooling rates of −0.1 to −4000 K/s and heating rates of 0.5 to 40 000 K/s possible. The instrument's resolution is 0.005 K and 0.01 K in the temperature range of 0.005 to 250 °C and −100 to 400 °C, respectively. The maximum sampling rate is 10 kHz. [261]. Due to small sample size required for high heating rates, no balancing during thermal degradation is possible.

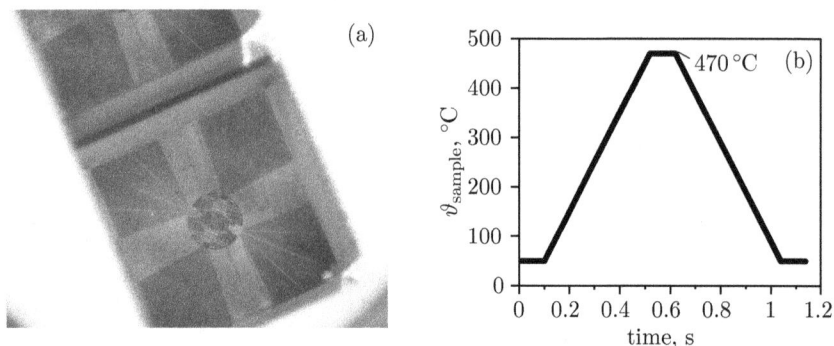

Figure 4.12: Flash-DSC: (a) MEMS chip sensor (b) temperature program

Scan procedure

The scanning procedure is based on an approach given by Cebe et al. [262]: Sample images are taken before and after scanning on an Olympus BX41 microscope with DCM 510 camera (10x lens). During scanning the sample environment is flushed with a flow rate of about 50 ml/min inert nitrogen. The ceramic support temperature is set to 298 K. The empty sensor is conditioned according to the manufacturer's procedure, by four or five heating and cooling cycles at ±2000 K/s to 450 K. Then, using a fine wire, a small sample particle is placed on the sample membrane of the sensor. If not denoted otherwise, the temperature program as depicted in Figure 4.12(b) is applied to the sample (repeatedly). The sample is kept at 50 °C for 0.1 s, is then heated with 1000 K/s to 470 °C, isothermally treated for 0.1 s and subsequently cooled with −1000 K/s to 50 °C and kept at 50 °C for 0.1 s. This cycle is repeated for most samples for two to four times.

[‖]Measured at Polymer Physics Group of Rostock University, Germany

4.3 Evaluation of experiments

For the evaluation of the pyrolysis experiments, the product yields of permanent gas (components), pyrolysis oil (components) and char are calculated.

4.3.1 Permanent gas yield

During pyrolysis, the permanent gases carbon monoxide, carbon dioxide, diverse hydrocarbons, and hydrogen are formed. For determination of the gas yield, the formed mass flow of each gas component has to be calculated from the measured gas concentrations. The calculation for CO, CO_2, and hydrocarbons C_mH_n (measured as CH_4) is carried out with the online measured concentrations. Both the hydrogen gas content as well as the hydrocarbon composition is calculated from the gas sample bag concentrations. For calculation, two positions in the process flowsheet as shown in Figure 4.13 are of interest. Firstly, the effluent of the CFB pyrolysis reactor before the side stream sampling PS, which equals the reactors total fluid effluent flow. Secondly, the influent of the online analyzers behind the cooler GA. For each stream in the system the sum of molar fractions must follow the equation (with varying molar fractions):

$$\sum y_i = 1 \qquad \text{with } i = \text{N}_2,\ \text{H}_2\text{O},\ \text{oil},\ \text{CO},\ \text{CO}_2,\ \text{C}_m\text{H}_n \text{ and } \text{H}_2 \qquad . \qquad (4.3)$$

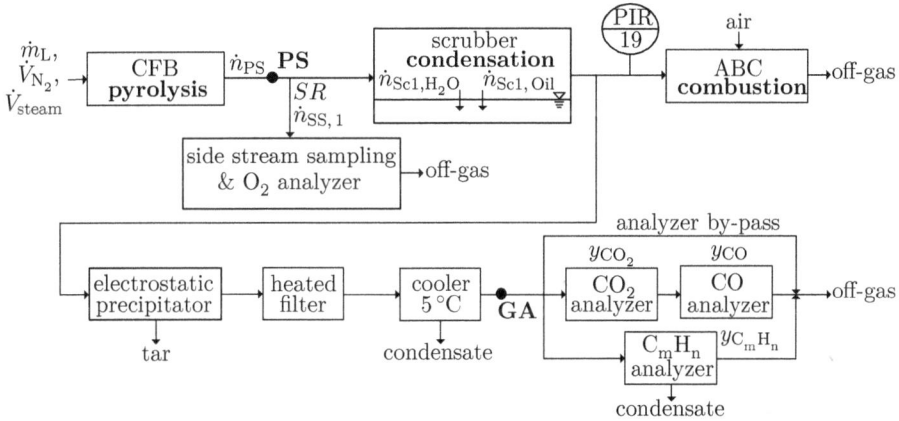

Figure 4.13: Simplified balancing scheme for permanent gas

At the point GA, it can be assumed that no pyrolysis oil remains in the stream as the oil is condensed in the scrubber Sc1 and the stream is subsequently cleaned in the electrostatic precipitator and heated filter. The molar fraction of CO, CO_2, and C_mH_n corresponds to the online measured values. As water is used as washing medium, much water is in the effluent of the scrubber Sc1. Therefore, behind the cooler, the gas is saturated with water at the cooler temperature of $5\,^\circ\text{C}$. The molar fraction of water can thus be calculated from the vapor pressure

$$y_{\text{GA},\text{H}_2\text{O}} = \frac{p^S_{\text{GA},\text{H}_2\text{O}}}{p_{\text{GA}}} \approx \frac{p^S_{\text{GA},\text{H}_2\text{O}}}{p_{19}} \qquad . \qquad (4.4)$$

Herein, the pressure p_{GA} is approximated by the absolute pressure at PIR19 p_{19}. For the water vapor pressure the Antoine-equation [263] does apply:

$$p^S_{GA,H_2O} \text{ in kPa} = 10^{7.31549 - \frac{1794.88}{-34.764 + T \text{ in K}}} \quad . \tag{4.5}$$

The hydrogen gas content is either considered by

$$y_{GA,H_2} = (y_{GA,CO} + y_{GA,CO_2} + y_{GA,C_mH_n}) \cdot \frac{y_{GB,H_2}}{y_{GB,CO} + y_{GB,CO_2} + y_{GB,C_mH_n}} \tag{4.6}$$

with $y_{GA,i}$ the molar fractions at the online analyzers and $y_{GB,i}$ the fractions in the corresponding gas bag or omitted if no gas bag is available. With the molar fractions of the permanent gases and water known, the molar fraction of N_2 can be calculated by Eq. 4.3. Although the concentrations of water and pyrolysis oil do change it can expected that the ratio of permanent gases and N_2 stays constant in the whole system. Hence, the pyrolysis reactors effluent permanent gas molar flow rates $\dot{n}_{PS,i}$ satisfy the following equation:

$$\dot{n}_{PS,i} = \frac{y_{GA,i}}{y_{GA,N_2}} \cdot \dot{n}_{PS,N_2} \quad \text{with } i = CO, CO_2, C_mH_n \text{ and } H_2 \quad . \tag{4.7}$$

The unknown molar flow rate $\dot{n}_{PS,N_2}$ of N_2 can be calculated from the sum of volume flows of fluidizing gases (riser and syphon) and feeding gas $\dot{V}_{N_2}$. Using the ideal gas law provides:

$$\dot{n}_{PS,N_2} = \frac{p \cdot \dot{V}_{N_2}}{R \cdot T} \tag{4.8}$$

with $\dot{V}_{N_2}$, p and T the volume flow, pressure, and temperature at standard conditions. The volume flow of $\dot{V}_{N_2}$ is determined from the rotameter scale together with the pressure (5 bar) and temperature ($\sim$10 °C) at the rotameter. From the molar flows the mass flows can easily be calculated by multiplication with the gas components molar mass:

$$\dot{m}_{PS,i} = \dot{n}_{PS,i} \cdot M_i \quad \text{with } i = CO, CO_2, C_mH_n \text{ and } H_2 \quad . \tag{4.9}$$

With the lignin mass flow $\dot{m}_L$ the gaseous product yield is given by

$$Y_{g,i} = \frac{\dot{m}_{PS,i}}{\dot{m}_L \cdot (1 - w_{H_2O,L})} \quad \text{with } i = CO, CO_2, C_mH_n \text{ and } H_2 \quad , \tag{4.10}$$

the overall gas yield thus is

$$Y_g = \sum_i Y_{g,i} \quad \text{with } i = CO, CO_2, C_mH_n \text{ and } H_2 \quad . \tag{4.11}$$

4.3.2 Pyrolysis oil yield

The simplified flowsheet for balancing of pyrolysis oil is shown in Figure 4.14. The determination of pyrolysis oil yield is obtained from the side stream sampling. Thus, as only a portion of the pyrolysis oil is condensed in the sample train, the split rate SR is necessary to extrapolate from the side stream SS to the overall yield at PS. The calculation of the split rate SR is based on the following assumptions:

- The molar flow out of the sample train at SS2 contains N_2, permanent gases, isopropanol, and H_2O.

- The temperature and relative pressure at FI40 are equal to T26 and P27, respectively.

- The flow is saturated with isopropanol and H_2O at T26.

- Due to high sampling temperature and the multiple changes of sample trains during an experiment, a leakage in the glass-glass connection of the impinger bottles cannot always be avoided thoroughly. Thus, a possible leakage of air has to be considered and can be calculated from the oxygen content measured behind the sample train.

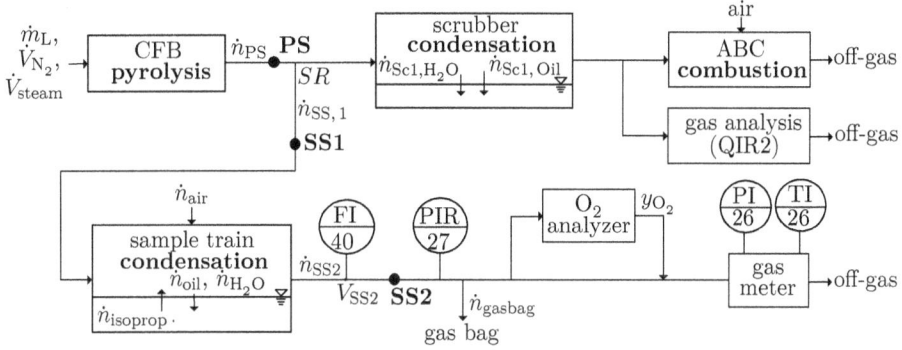

Figure 4.14: Simplified balancing scheme for pyrolysis oil

As the molar flow rate of permanent gases does not change between the point SS1 and SS2 (cf. flowsheet) and the N_2 flow can be corrected by the known oxygen content it holds

$$SR = \frac{\dot{n}_{SS2,CO} + \dot{n}_{SS2,CO_2} + \dot{n}_{SS2,C_mH_n} + \dot{n}_{SS2,H_2} + \dot{n}_{SS2,N_2}}{\dot{n}_{PS,CO} + \dot{n}_{PS,CO_2} + \dot{n}_{PS,C_mH_n} + \dot{n}_{PS,H_2} + \dot{n}_{PS,N_2}} \ . \tag{4.12}$$

The numerator can be obtained from the isopropanol, water, and air molar fraction as well as the total molar flow at SS2:

$$\dot{n}_{SS2,CO} + \dot{n}_{SS2,CO_2} + \dot{n}_{SS2,C_mH_n} + \dot{n}_{SS2,H_2} + \dot{n}_{SS2,N_2} = \dot{n}_{SS2}$$
$$\cdot (1 - y_{SS2,\,isopopanol} - y_{SS2,H_2O} - y_{SS2,\,air}) \ . \tag{4.13}$$

The molar flow rate is calculated from the volume flow at the rotameter FI40 and the ideal gas law (Eq. 4.8). The molar fraction of water with temperature T26 is analogously obtained from Eq. 4.4 and the Antoine correlation for water Eq. 4.5. The same holds for the molar fraction of isopropanol with the Antoine correlation [263]

$$p^S_{SS2,\,isopopanol} \text{ in kPa} = 10^{7.24268 - \frac{1580.92}{-53.54 + T \text{ in K}}} \ . \tag{4.14}$$

The measured oxygen content together with the volume fraction of oxygen in air yields

$$y_{SS2,\,air} = \frac{y_{SS2,O_2}}{0.21} \ . \tag{4.15}$$

With the split rate known, the total mass flow of pyrolysis oil can be extrapolated from the condensed oil in the sample train:

$$\dot{m}_{\text{PS,oil}} = \frac{\dot{m}_{\text{SS1,oil}}}{SR} \qquad . \tag{4.16}$$

For the pyrolysis oil mass flow in the side stream, it holds

$$\dot{m}_{\text{SS1,oil}} = \frac{m_{\text{tar1}}}{\Delta t_{\text{SS}}} \tag{4.17}$$

with m_{tar1} the mass of oil weighed after rotary evaporation (cf. path 2b in Figure 4.10) and the sampling duration Δt_{SS}. With the lignin mass flow $\dot{m}_L$ the liquid product yield is given by

$$Y_{\text{oil}} = \frac{\dot{m}_{\text{PS,oil}}}{\dot{m}_L \cdot (1 - w_{\text{H}_2\text{O,L}})} \qquad . \tag{4.18}$$

The mass of a pyrolysis oil component in a side stream sample $m_{i,\text{oil, SS}}$ is obtained as explained in Section 4.2.2 "Gas chromatography (GC)". Together with the split rate SR, lignin mass flow during sampling $\dot{m}_{\text{L, SS}}$, its water content $w_{\text{H}_2\text{O,L}}$ and the sampling duration Δt_{SS} the yield of component $Y_{\text{oil},i}$ can be calculated:

$$Y_{\text{oil},i} = \frac{m_{i,\text{oil, SS}}}{\dot{m}_{\text{L, SS}} \cdot (1 - w_{\text{H}_2\text{O,L}}) \cdot \Delta t_{\text{SS}}} \qquad . \tag{4.19}$$

4.3.3 Char yield

For determination of the char yield, the overall char mass produced during the pyrolysis experiment m_{char} is to be determined:

$$m_{\text{char}} = m_{\text{char, BM}} + m_{\text{char, C2}} + m_{\text{char, comb}} + m_{\text{ash, L}} \qquad . \tag{4.20}$$

Herein, $m_{\text{char, BM}}$ is the mass of char in the bed material, $m_{\text{char, C2}}$ the char mass in the secondary cyclone material, $m_{\text{char, comb}}$ the char mass that remained in the reactor, which is determined through incineration before the next experiment, and $m_{\text{ash, L}}$ the mass of ash in the fed lignin. Endmost holds for the assumption that all mineral matter remains in the solid residue during pyrolysis. The mass of inert free char in the bed material is obtained from the total mass of bed material, drained from the CFB system after experiment m_{BM} and its water $w_{\text{H}_2\text{O,BM}}$ and inert $w_{\text{I, BM}}$ content:

$$m_{\text{char, BM}} = m_{\text{BM}} \cdot (1 - w_{\text{H}_2\text{O,BM}} - w_{\text{I, BM}}) \qquad . \tag{4.21}$$

The mass of inert free char in the secondary cyclone material is obtained analogously to Eq. (4.21). It holds

$$m_{\text{char, C2}} = m_{\text{C2}} \cdot (1 - w_{\text{H}_2\text{O,C2}} - w_{\text{I, C2}}) \qquad . \tag{4.22}$$

The amount of char that can not be drained from the CFB system, is calculated from the measured CO and CO_2 volume fraction during heating up for the following experiment. The char incinerates due to oxidation, which starts at about $400\,^\circ C$. It holds

$$m_{char,comb} = \frac{\bar{V}_{air} \cdot \Delta t_{comb}}{1 - 0.5 \cdot \bar{\phi}_{CO}} \cdot \left(\bar{\phi}_{CO} \cdot \rho_{CO} \cdot \frac{M_C}{M_{CO}} + \bar{\phi}_{CO_2} \cdot \rho_{CO_2} \cdot \frac{M_C}{M_{CO_2}} \right) \quad (4.23)$$
$$\cdot (1 + f_H + f_N + f_S + f_O) \quad .$$

Herein, $\bar{V}_{air}$ is the average air volume flow at standard conditions during the combustion duration Δt_{comb}, $\bar{\phi}_{CO}$ and $\bar{\phi}_{CO_2}$ the volume fractions of CO and CO_2 in the flue gas, respectively. Furthermore, $\rho_{CO} = 1.23\,kg/m^3$ and $\rho_{CO_2} = 1.95\,kg/m^3$ are the densities of the combustion products (at standard conditions), M_i the molar masses of $i = $ C, CO, and CO_2 and f_i the solids loading of carbon with the other char components H, N, S, and O. As the ash content in the bed and secondary cyclone material samples can not be measured directly (cf. section 4.2.2 – Elemental analysis) it can only be approximated by the ash content in the fed lignin $w_{A,L}$. With the assumption that all mineral matter remains in the char (which is not entirely correct, cf. section 2.2.4.1) the mass of ash in the char is approximately determined by Eq. (4.24):

$$m_{ash,L} = m_L \cdot w_{A,L} \quad . \quad (4.24)$$

The total char yield (including ash) depends on the total amount of fed lignin m_L, and the total mass of produced char (Eq. (4.20)). It holds:

$$Y_{char} = \frac{m_{char}}{m_L \cdot (1 - w_{H_2O,L})} \quad . \quad (4.25)$$

4.4 Used material

In this section, the used fluidized bed heat carrier quartz sand, Kraft, and hydrolysis lignin are characterized. The lignins are additionally compared by thermal characterization to reference biomasses (cellulose, xylan, organosolv and soda lignin).

4.4.1 Lignin

Preparation of lignins

Kraft lignin
The Kraft lignin used for pyrolysis experiments was produced by Kraft pulping of softwood chips with subsequent separation from the black liquor by the LignoBoost process (Innventia AB, Sweden). Pulping conditions are cooking in an aqueous solution with NaOH and Na_2S at about 165 to $175\,^\circ C$ for a digestion time at the maximum temperature of 1 to 2 h [264]. The lignin molecules are cleaved, dissolve and are separated from the pulp. The lignin is then precipitated from the black liquor by acidification, preferably with CO_2, and filtered before drying [265, 266]. The primary particle size distribution as received is depicted in Fig. 4.15a (lignin). Due to the small particle size, which is leading to bridge forming in the lignin hopper, the Kraft lignin is agglomerated with methyl cellulose to a particle size of 0.5 to 2 mm (cf. aggl. lignin in 4.15a). A picture of the agglomerated lignin is shown in Fig. 4.15b.

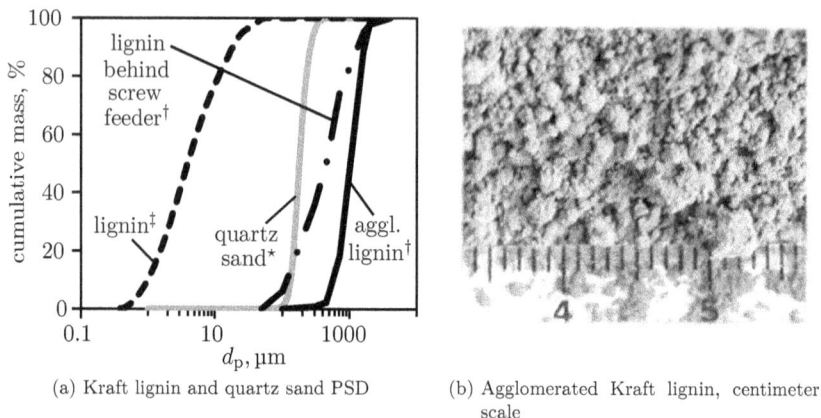

(a) Kraft lignin and quartz sand PSD

(b) Agglomerated Kraft lignin, centimeter scale

Figure 4.15: Cumulative mass distribution (compared to quartz sand) and picture of Kraft lignin

measured with: *Camsizer XT, †sieve analysis, ‡Beckman Coulter

Hydrolysis lignin

As second feed material for pyrolysis hydrolysis lignin is used. It was produced at the Institute of Thermal and Separation Processes, Hamburg University of Technology, which acquired consolidated knowledge in the field of biomass pretreatment in the past years [13, 41, 267–272]. Straw is hydrolyzed in a 40 l single high pressure fixed bed. A batch of 11 kg wet straw is cooked in the reactor at 200 °C and 40 bar. The fixed bed is passed through upstream by a water mass flow of 180 kg/h for 30 min. In this process step the hemicellulose is dissolved in the water, while cellulose and lignin remain in the beds material. Successively, the bed is transferred into a second reactor, where it is further enzymatically hydrolyzed for 72 h at 50 °C, 1 atm, a pH of 5 and an enzyme concentration of 15 fpu (filter paper units). Cellulose is degraded to glucose during this process step by a cellulase mixture (Cellic® CTec2 of Novozyme A/S, Denmark) which contains also β–glucosidases and hemicellulases. Glucose dissolves and the remaining hydrolysis residue is decanted from the substrate and dried at 50 °C creating a hard solid cake like structure. Cake pieces (10 x 10 x 5 cm^3) are then ground by a jaw crusher (Fritsch GmbH, "Pulverisette 1") and a cone mill (W. Feddeler GmbH & Co. Laborbedarf, "Laborkegelbrecher FE 10") and sieved to a particle size of < 1.8 mm resulting in the feed material used for pyrolysis (particle size distribution cf. Figure 4.16a). A picture of the material is shown in Fig. 4.16b.

Reference material for thermal characterization

As reference materials for thermal analysis cellulose, xylan, organosolv and soda lignin are used. The cellulose particles of type Sigmacell Cellulose S3504 have been purchased from Sigma-Aldrich. Xylan from birch wood serves as hemicellulose model compound and is obtained from Sigma-Aldrich as well. It has the chemical structure Poly(β-D-xylopyranose[1→4]). Both organosolv and soda lignin were produced at the Thünen Institute of Wood Research, Germany from the same wheat straw as the hydrolysis lignin. The soda pulping was carried out in a laboratory scale reactor with 16% NaOH at 160 °C. The liquor residence time was 60 min, the liquor ratio 40. The lignin was then precipitated with H_2SO_4. Analogously, the organosolv lignin was obtained by cooking the wheat straw with an ethanol-water mixture of 1:1 in a laboratory scale reactor with 2% H_2SO_4 at 180 °C.

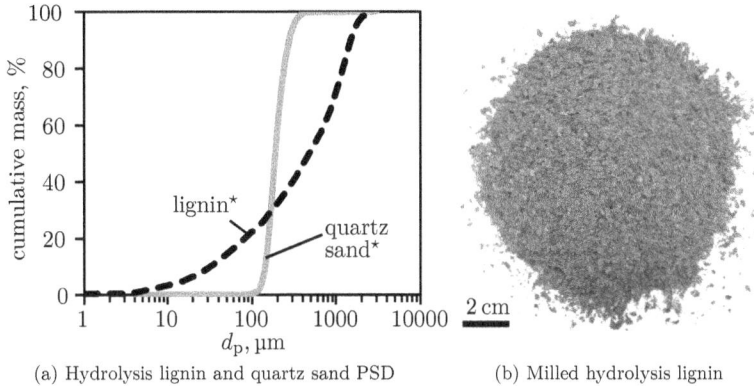

(a) Hydrolysis lignin and quartz sand PSD   (b) Milled hydrolysis lignin

Figure 4.16: Cumulative mass distribution (compared to quartz sand) and picture of hydrolysis lignin

measured with: *Camsizer XT

The residence time was 2 h, the liquor ratio 4. Precipitation is achieved by dilution with water. The pulping residues were drained by by centrifuging in 2 steps and dried in two steps in air at 55 °C and with P_2O_2 at 40 °C. Finally, the lignins were milled.

General lignin characterization

In the screw feeder, comminution takes place with a resulting particle size distribution for Kraft lignin (Fig. 4.15a, lignin after screw feeder). For the hydrolysis lignin, which is much harder than the Kraft lignin agglomerates, less size reduction occurs when it is fed through the pyrolysis plants screw feeder. The resulting cumulative particle size distribution is depicted in Fig. 4.16a. The Sauter diameters are provided in Table 4.2. SEM images for lignin fines are depicted in Figure 4.17. It can be observed that the Kraft lignin is more homogenous in shape (Fig. 4.17a), whereas the hydrolysis lignin also contains twig like particles (cf. Fig. 4.17b). This morphology can be explained by incomplete conversion in the hydrolysis process (cf. above) resulting in considerable amounts of cellulose ($\sim$ 39 wt.-%) and hemicellulose ($\sim$ 5 wt.-%) in the hydrolysis lignin. It only has a purity of 49 wt.-% compared to Kraft lignin with about 95.5 wt.-% as well as 89.2 wt.-% and 61.3 wt.-% for the reference materials organosolv and soda lignin, respectively (cf. Table 4.3). Table 4.4 lists the further composition of the lignins. Notable are the differences in carbon and oxygen content (due to different carbohydrate content) and the high ash content of hydrolysis and soda lignin. The latter originates from the higher ash content in straw (8.2 wt.-%) compared to woody biomass (e.g. 0.3 wt.-% for spruce wood) [14] and differences in the pretreatment processes. The lignins also possess different inorganics composition. Part of these differences might be accounted for by the used pulping and recovery chemicals in the Kraft process: S, Na, and Ca [32]. Furthermore, lignocellulosic biomass is known to have larger concentrations of potassium and calcium, whereas sodium, magnesium and other metals are normally contained in a lower amount [273]. The molar mass averages for the lignins are given in Table 4.5. They span from about 1500 to 2250 g/mol and from 5300 to 14 800 g/mol for M_n and M_w, respectively. The polydispersity index ranges from 3.6 to 8.9 showing the samples have a different degree of heterogeneity of molar weight distribution. The lignin densities are summarized in Table 4.2, showing a slightly lower density for Kraft lignin. The material densities ρ_m of the reference biomass samples are 1480, 1500, 1290, and $1480 \, \text{kg/m}^3$ for cellulose, xylan, organosolv, and soda lignin, respectively.

Table 4.2: Densities, Sauter diameter and minimal fluidization velocity of used materials

		Kraft lignin			Hydrolysis lignin, fed	Quartz sand
		fines	aggl.	fed		
ρ_{m}	kg/m^3	1317	1317	1317	1436	2604
ρ_{a}	kg/m^3	1216	1216	1216	1320	2590
ρ_{bulk}	kg/m^3	n.d.	449	n.d.	565	1325
d_{sauter}	µm	2.5	944	301	82	175
$u_{\text{mf}}{}^\dagger$	m/s	n.d.	n.d.	n.d.	n.d.	0.0213

†measured in air at 1 bar and 20 °C

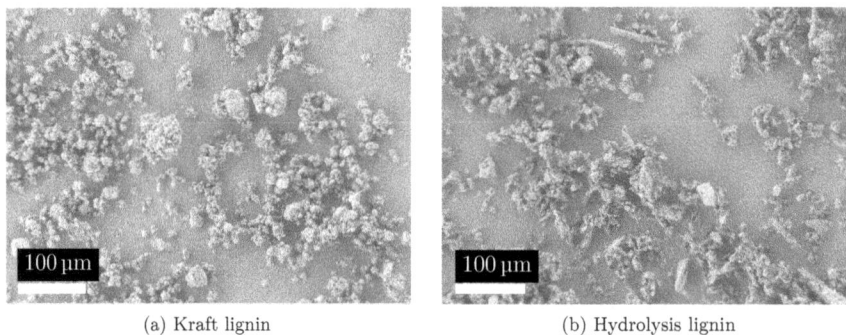

(a) Kraft lignin   (b) Hydrolysis lignin

Figure 4.17: SEM images of lignin fines

Thermal characterization

Conventional DSC/TGA measurements are carried out at heating rates of 0.02 to 300 K/min. But the heating rate in fluidized beds – especially for small particles – can reach up to 10^5 K/s [50, 51] and for average conversions about 1100 to 8 K/s [131]. Thus, conventional DSC/TGA measurements do not reflect the actual heating rate behavior. As only few investigations on biomass component behavior at these high heating rates have been reported in literature, it is of interest to analyze the thermal treatment at high heating rates. The progress in development of analytical equipment has recently (in 2011) brought about the commercialization of the Flash-DSC 1. With this equipment, the characterization of thermal behavior is attempted here. Preliminary experiments have shown that the decomposition of lignin samples also takes place at a heating rate of 10 000 K/s, but the material transitions can best be observed at a heating rate of 1000 K/s. Furthermore, 1000 K/s are close to possible heating rates in fluidized bed systems ([50, 51]) and thus selected for sample analysis.

The Flash-DSC diagrams (e.g. in Figure 4.18(a)) show curves for the heat flow into the sample (heating) and out of the sample (cooling). The heating curve is the lower part of a heating and cooling cycle, followed by a vertical line, which is representing the isothermal treatment between heating and cooling. Successively, the cooling curve (the upper part of the heating and cooling cycle) and a further vertical isothermal complete

Table 4.3: Composition of Kraft and hydrolysis lignin in comparison to reference biomass samples

| | | component wt.-%, dry | | | | |
| | | lignin | | | cellulose | xylan |
	Kraft	hydrolysis	organosolv	soda		
xylose	0.46	4.48	1.96	12.26	0.01	76.42
glucose	0.23	38.38	0.78	1.68	0.97	0.65
mannose	0.11	0.35	0.08	0.07	0.02	0.00
galactose	0.67	0.25	0.12	0.33	0.00	0.21
arabinose	0.27	0.30	0.20	2.31	0.00	0.00
rhamnose	0.00	0.05	0.01	0.04	0.00	0.25
sugars total	1.74	43.80	3.15	16.69	1.00	77.54
lignin solid residue	90.67	46.93	89.20	61.30	0.00	0.43
lignin acid soluble	4.77	1.93	n.d.	n.d.	n.d.	n.d.
lignin total	95.43	48.85	89.20	61.30	0.00	0.43

The difference to 100% are not detected oils, proteins, waxes and losses in the analytical hydrolysis

Table 4.4: Ultimate, partial proximate, and inorganic content analyses as well as higher heating value of Kraft and hydrolysis lignin in comparison to reference biomass samples

	Kraft lignin	Hydrolysis lignin	Organosolv lignin	Soda lignin	Cellulose	Xylan
Ultimate analysis in wt.-%, dry bases						
C	65.9	48.0	66.0	46.4	43.8	43.6
H	6.0	6.3	6.4	5.9	7.4	7.6
O†	25.0	33.3	24.7	24.8	48.6	48.6
N	0.2	0.8	2.1	0.5	< 0.1	< 0.1
S	1.8	0.3	0.2	2.4	0.2	0.2
Partial proximate analysis in wt.-%						
Moisture	2.2	3.9	2.4	5.8	4.3	11.3
Ash, dry bases	1.1	11.3	0.6	20.0	0	0
HHV‡ in MJ/kg, dry bases						
	27.7	20.5	28.0	20.4	19.0	19.1
Inorganic content analysis, dry bases						
Na in g/kg	2.8	1	n.d.	n.d.	n.d.	n.d.
K in g/kg	0.7	< 4	n.d.	n.d.	n.d.	n.d.
Ca in g/kg	0.6	4.9	n.d.	n.d.	n.d.	n.d.
Cu in mg/kg	< 5	11.6	n.d.	n.d.	n.d.	n.d.
Mg in mg/kg	55.3	700	n.d.	n.d.	n.d.	n.d.
Al in g/kg	0.2	0.3	n.d.	n.d.	n.d.	n.d.
Fe in mg/kg	33.2	545.0	n.d.	n.d.	n.d.	n.d.
Mn in mg/kg	56.8	25.5	n.d.	n.d.	n.d.	n.d.
Ni in mg/kg	< 50	18.6	n.d.	n.d.	n.d.	n.d.
Zn in mg/kg	52.2	18.7	n.d.	n.d.	n.d.	n.d.
Li in mg/kg	< 5	< 50	n.d.	n.d.	n.d.	n.d.

†by difference, ‡HHV correlation by Eq. 4.1

Table 4.5: Molar mass (number M_n and weight M_w average) and polydispersity index PD for Kraft, hydrolysis and reference lignins

	M_n	M_w	PD
	g/mol	g/mol	(-)
Kraft lignin	2257	12949	5.7
hydrolysis lignin (straw)[†]	1670	14821	8.9
organosolv lignin (straw)	1462	5272	3.6
soda lignin (straw)	2007	8157	4.1

[†]presumably only lignin fraction dissolved in DMSO

the cycle. The hysteresis between the heating and cooling curves is a measure of sample weight. Thus, as the sample loses weight through drying and devolatilization, the distance between the curves of heating and cooling heat flux is reduced. The curves converge somewhat on every cycle. For visual insight into what changes a sample undergoes, microscopic pictures before and after thermal treatment are taken. The sample particles placed on the measuring chip were measured in size (by scaling of particles in microscopic images) and the mass was approximated by an ellipsoid volume and the apparent material density. Sample particle masses ranged from 0.2 to 10 µg.

Lignins are known to be fundamentally amorphous showing local mode relaxation, glass transition and decomposition with rising treatment temperature. Further, it is believed that lignin shows no first order thermodynamic phase transitions, indicating that solid lignin is either in glassy or rubbery state [274]. When Kraft lignin is heated with 1000 K/s up to 470 °C drying, glass transition, softening, melting and pyrolysis reactions can be observed. Glass transition can be seen at 170 to 190 °C in contrast to literature data. Therein, glass transition temperatures ϑ_g are reported for milled wood and Kraft lignin in the range of 110 to 180 °C and 100 to 174 °C, respectively [46, 47, 274, 275]. Softening and melting starts in a temperature range of 215 to 240 °C and continues to about 285 °C followed by reactions from about 315 °C. The reactions continue until the temperature declines below 315 °C in cooling (solid line in Figure 4.18(a)). In the second cycle (dotted line) reaction is recurring at higher temperature of 435 to 455 °C during heating.

Figure 4.18(b) shows the behavior at the same conditions for a hydrolysis lignin particle. Neither a phase change nor a distinct reaction region can be observed. But it can be seen that, as the heating and cooling curves draw nearer to each other, some mass loss occurs. Also, the pictures of the particle before and after thermal treatment show that no phase change (melting) took place. Furthermore, the particle shrank by about 11 area-% and its color became darker, indicating a mass loss. This mass loss might be attributed to drying and a partial reaction of cellulose, hemicellulose, and lignin (cf. composition of hydrolysis lignin in Table 4.3), which together react in a broad temperature range of 200 to 500 °C (cf. Figure 2.7).

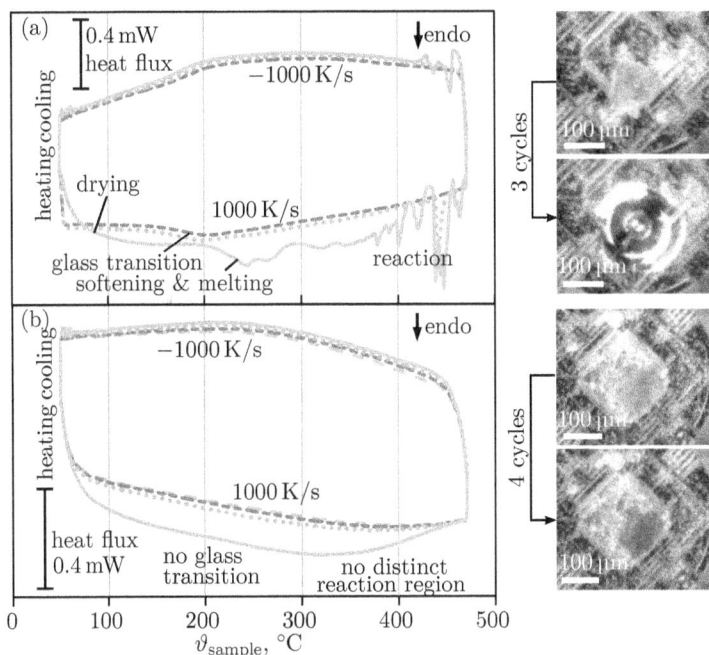

Figure 4.18: Flash-DSC heat flow rates to (a) Kraft and (b) hydrolysis lignin particles

It is obvious that the thermal behavior is fundamentally different for the two lignins. To shed some light on the reason for this difference cellulose, xylan as a model compound for hemicellulose, organosolv, and soda lignin are investigated. Cellulose did not react at higher heating rates, which could be due to bad heat transfer from the chip to the particle, induced by the fact that the particle does not melt (small contact area between chip and particle). Thus, cellulose was investigated at $30\,\mathrm{K/min}$ matching a heating rate of conventional DSC (Figure 4.19(a)). No phase transition can be seen but the particle reacts almost completely in one cycle, leaving only a small residue. Additionally, the shape of the residue indicates that no melting did take place. The reaction is first endothermic and then exothermic and is complete at about $400\,^{\circ}\mathrm{C}$ on cooling. Xylan in Figure 4.19(b) shows also no melting and degradation starts at $400\,^{\circ}\mathrm{C}$ during first heating, is then interrupted when cooled below about $360\,^{\circ}\mathrm{C}$ and continues in the second cycle from $380\,^{\circ}\mathrm{C}$ (heating) to stop at $385\,^{\circ}\mathrm{C}$ during second cooling. Further some minor reactions take place in the third cycle. The xylan sample retains a particulate structure, which darkens in color with progressing degradation to form a black residue after four cycles. In a conventional DSC xylan degrades at the lowest temperature, then cellulose, and lignin degradation has its maximum at the highest temperature but reacts over the broadest temperature range (cf. Figure 2.7). Contrary to this behavior, here cellulose reacts not at all at the high heating rates, whereas lignin does and xylan starts to react at higher temperature of more than $400\,^{\circ}\mathrm{C}$. This finding could be due to the difference in heat transfer at the contact interface between chip membrane and sample. The heat transfer for a solid sample is inferior compared to molten sample due to smaller contact area. For further investigation, this heat transfer could maybe be enhanced by using a thin layer of oil, with known properties, between the sample and the membrane, making a good wet contact [276].

Figures 4.19(c) and 4.19(d) show the Flash-DSC measurement curves and microscopic pictures for soda and organosolv lignin. The behavior of soda is similar to that of hydrolysis and that of organosolv similar to Kraft lignin. Soda lignin, like hydrolysis lignin does not melt. The only difference for soda lignin is a bigger shrinkage of about 45% by area, which can partly be explained by some degradation reactions above 400 °C. Organosolv lignin shows similar melting behavior compared to Kraft lignin. The endothermic melting peak is greater, probably mainly because of bigger particle size and the degradation reactions starting more vigorous at ∼350 °C. This observation is in good agreement with the results from TGA analysis (10 K/min) of wheat straw derived organosolv lignin [29], which shows a maximum thermal degradation around 365 °C. Furthermore, the formed droplet contains several bubbles.

It can be expected, that the hottest reaction zone in the sample is the contact interface between the particle or formed droplet and the chip membrane. Thus, in case of a molten sample (droplet), gaseous reaction products form bubbles in the proximity of the hot chip surface. The bubbles formed can be observed in each microscopic droplet picture in Figures 4.18(a) and 4.19(d) for Kraft and organosolv lignin, respectively. It can be seen that one bubble has formed right in the middle of the intermediate product droplet from Kraft lignin. For organosolv lignin, multiple bubbles can be seen, which either are released from the droplet or coalesce after repeated thermal treatment. The fluctuations of the measured heat flux are caused by the reaction heat needed for sample decomposition. But it should also be considered that the formation of bubbles might induce further fluctuations in heat flux. As bubbles are formed and grow due to the ongoing reactions, bubble movement, coalescence, and release create a change in the chip-sample interface. Thus, as a changing fraction of the sample at the chip-sample interface is either gaseous bubble or liquid droplet, a change in heat transfer is caused. If for example, a bubble which is in contact with the membrane surface is released from the droplet rapidly, the bubble detachment from the membrane surface will presumably, due to the higher solid-liquid heat transfer at the now liquid-membrane surface, instantly increase the heat flux to the sample.

To summarize the results, it can be concluded that the difference in thermal behavior can be explained by the diverse sample composition. But, biomass origin as reason for deviation in thermal behavior can be eliminated due to the fact that hydrolysis and soda lignin (not melting), as well as organosolv lignin (melting) were produced from the same primary material (wheat straw). Furthermore, on the particle level also the particle size contrary to de Wild et al. [29] seems to be of minor importance as the small hydrolysis and soda lignin particles (size range of 50 to 100 µm) do not melt. Thus, it can be deduced, that a higher impurity of a lignin sample with cellulose and hemicellulose inhibits melting.

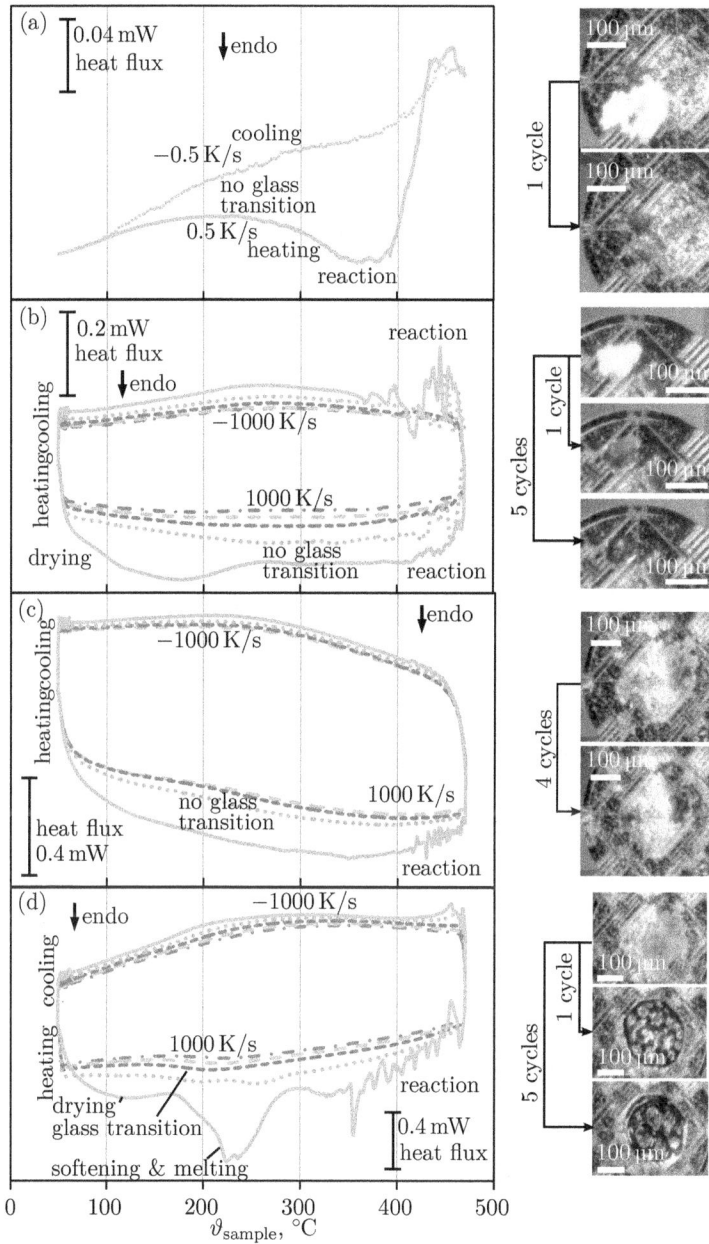

Figure 4.19: Flash-DSC heat flow rates to (a) cellulose, (b) xylan, (c) soda lignin, and (d) organosolv lignin particles

4.4.2 Quartz sand

The heat carrier for the circulating fluidized bed, quartz sand of type F36, is procured from Quarzwerke GmbH, Germany. The sand particle size distribution is shown in Figure 4.21. Figure 4.20 shows a SEM image of the quartz sand. Its composition is given in Table 4.6. Sauter diameter, densities, and minimal fluidization velocity are listed in Table 4.2.

Table 4.6: Quartz sand F36 composition from supplier data sheet [277]

SiO_2 wt.-%	Al_2O_3 wt.-%	Fe_2O_3 wt.-%	Loss on ignition[†] wt.-%
99.3	0.5	0.06	0.2

[†]according to DIN EN ISO 3262-1 at 1000 °C

Figure 4.20: SEM image of quartz sand

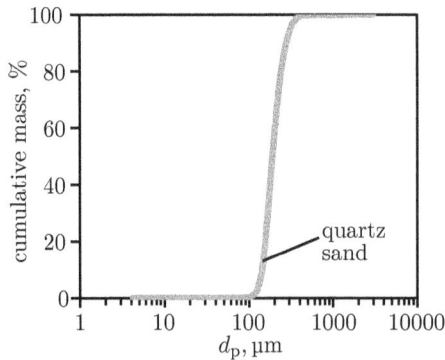

Figure 4.21: Cumulative mass distribution of quartz sand measured with Camsizer XT

4.5 Pyrolysis experiments

In Table 4.7 the operating conditions of the pyrolysis experiments of this work are listed. Kraft lignin was pyrolyzed in 14 and hydrolysis lignin in four experiments. For Kraft lignin, the temperature was varied between 550 to 700 °C in 50 K steps. The hydrolysis lignin pyrolysis temperature was varied in 100 K steps in the range of 500 to 700 °C. The experiments further varied in duration 47 to 162 min, lignin mass flow rate 1.34 to 3.34 kg/h for Kraft and 3.64 to 4.44 kg/h for hydrolysis lignin. The pyrolysis experiments for hydrolysis lignin were carried out without steam as fluidizing gas, while for most Kraft lignin experiments a mixture of steam and N_2 was used. Only during experiments 71, 72 and 89 the inert gas solely consisted of N_2. Although an influence of steam atmosphere exists (cf. Section 2.2.3) and has also been observed to a small extent in this work, it is not further considered in the discussion.

Table 4.7: List of CFB pyrolysis experiments and operating conditions

exp. no.	lignin type	ϑ °C	t_{feed} min	m_L kg	$\dot{m}_L$ kg/h	$\dot{m}_{N_2}$ kg/h	$\dot{m}_{H_2O}$ steam kg/h	u_{N_2} riser	u_{H_2O} riser	u_{N_2} feed	u_{N_2} syphon
57	KL	650	68.0	2.46	2.15	21	12	2.1	2.5	41.8	0.17
58	KL	650	32.2	0.94	1.76	21	12	2.1	2.5	40.5	0.16
61	KL	650	62.7	2.26	2.16	18	12	1.8	2.5	42.0	0.10
63	KL	600	51.9	1.84	2.13	20	12	1.8	2.4	42.0	0.16
64	KL	700	34.3	0.78	1.34	18	12	1.8	2.6	42.0	0.18
70	KL	650	36.0	2.00	3.34	21	12	2.1	2.5	39.4	0.17
71	KL	650	38.0	0.96	1.52	36	0	4.3	0.0	39.5	0.17
72	KL	650	50.2	1.43	1.71	36	0	4.3	0.0	39.4	0.17
73	KL	550	63.5	2.07	1.95	21	12	1.9	2.2	42.3	0.15
78	KL	650	116.5	3.93	2.02	22	12	2.3	2.5	42.3	0.17
81	KL	650	129.5	4.36	2.02	22	12	2.3	2.5	42.1	0.17
87	HL	600	67.5	4.31	3.83	35	0	3.7	0.0	41.8	0.21
88	HL	600	150.7	9.13	3.64	34	0	3.7	0.0	41.7	0.16
89	KL	650	162.4	6.07	2.24	34	0	4.0	0.0	41.9	0.17
90	HL	700	99.3	7.34	4.44	31	0	3.7	0.0	41.7	0.17
91	HL	500	95.2	6.34	4.00	36	0	3.6	0.0	41.8	0.14

[†]reference is cross sectional area of riser, feeding lance and syphon; prevailing temperature: ϑ in riser and syphon as well as 20 °C for feeding gas

5 Modeling of pyrolysis process

For simulation of the lignin pyrolysis process, a generalized semi-empirical fluidized bed reactor model was developed in the Aspen Custom Modeler® (ACM) language. The generalization has the advantage, that the model can be validated with diverse biomasses in different fluidized bed systems as limited data exists for lignin pyrolysis in fluidized beds. Also, the high impurity level of the hydrolysis lignin makes it necessary to consider substantial amounts of cellulose and hemicellulose in the feed material. The use of ACM, which is integrated into Aspen Plus® of Aspen Technology, Inc. has the further convenience that the ACM model can be imported and used in Aspen Plus® flowsheet models. Moreover, the comprehensive Aspen Properties Database® can be directly used for calculation of chemical media properties. The modeling of the fluidized bed reactor is based on previous work [278, 279] and the student projects carried out under the author's supervision (cf. Table A.4).

As pyrolysis is a multi-reaction process with numerous parallel and sequential heterogeneous solid-gas and homogeneous gas-gas phase reactions, an increase in gas volume flow occurs along the reactor height. Thus, the superficial gas velocity increases, which influences the fluid dynamics and solid entrainment. Also, the solid reactions (and granulation in the case of Kraft lignin) with the generation of char changes the solid properties, i.e. density and particle size distribution and thus again the fluid dynamics. The reaction kinetics depend on the local reactant concentrations, which reversely depend on the fluid dynamics. Therefore, the reactor model subdivisions (cf. Figure 5.1) have to be solved altogether. Solving is achieved by the integrated compiler and solver of the Aspen Custom Modeler® (ACM) program. The Aspen Custom Modeler® works object and equation-oriented, i.e. sequence and form of the equations do not matter. The equations are stated – in contrast to traditional programming languages – acausally, without the necessity to solve the set of DAE equations for the computed variable manually. The solution is obtained by the ACM® through discretization of the reactor volume in a number of n_{total} volume increments. The number of discretization elements n_{total} can be tuned manually before calculation. For discretization, the backwards finite difference method of first order (BFD1) was used.

The fluid dynamics of the fluidized bed reactor consider the reactor geometry, different zones, i.e. a dense zone with bubble and suspension phase and a dilute phase. Furthermore, the gas and solid fractions, velocities and residence times in each phase are calculated with rising reactor height. At the reactor outlet, the solids entrainment is calculated. The reaction scheme contains parallel reactions for cellulose, hemicellulose, and lignin to pyrolysis oil, gas, and char with the associated reaction rate constants. Mass transfer occurs by means of prevailing concentration gradients of gases between different phases and the change of moles due to reaction. As the lignin/ biomass reacts and char is formed the solid density and particle size change (mono-sized distribution) is determined. In case of pure (Kraft) lignin, this change in particle size is considered as granulation process with calculation of char layer growth. The mass balances integrate all model parts so that the concentration along the reactor height is obtained.

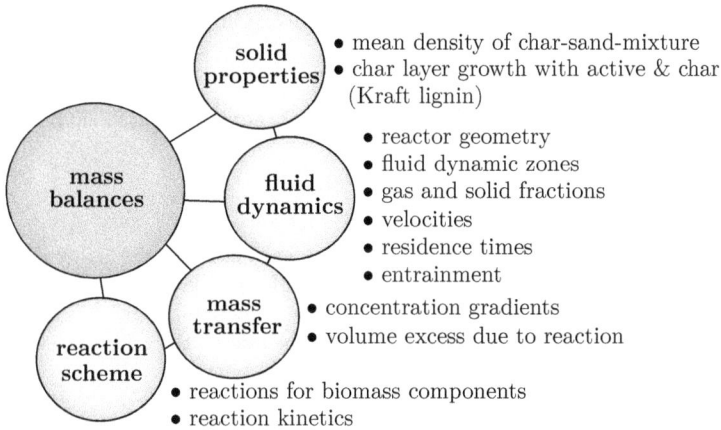

Figure 5.1: Model structure for lignocellulosic biomass pyrolysis in a fluidized bed reactor

5.1 Reactor fluid dynamics

For description of the fluid dynamics a semi-empirical model for fluidized beds, that is divided into two distinct zones, is used. The two zones are the dense bottom zone and the dilute upper zone. The dense bottom zone itself is divided into two phases according to a model by Werther and Wein [280] and Werther and Hartge [281, 282]. The two-phase model considers the coexistence of a bubble and a suspension phase. The bubble phase is assumed to be solids free and the gas to flow in plug flow. The suspension phase consists of both solid particles and gas. The particles are assumed to be ideally mixed, while the gas passes this phase also in plug flow. The dilute upper zone consists of a suspension phase only. Both gas and solids pass through this phase in plug flow. The solids fraction exponentially decreases with rising height. Virtual formation of clusters [106] and the division in lean upwards flowing (core) and dense reverse flowing (annulus) phase [283] are neglected with the purpose to develop a model that gives reasonable results for pyrolysis yields in a wide range of fluidized bed regimes. The division of the reactor volume into dense bottom and dilute upper zone occurs according to the persisting operation parameters. The calculation of the solids mass in each zone and the height of the inter-zone interphase is carried out by solving the overall mass balances. The total bed mass and reactor height are fixed. Summarized, the following assumptions are taken for model derivation:

1. division of fluidized bed into dense bottom and dilute upper zone

2. division of dense bottom zone in bubble and suspension phase

3. solids in bottom zone's suspension phase are ideally mixed

4. plug flow for gas phase in dense bottom zone and for gas and solids phase in dilute upper zone

5. dilute upper zone has exponentially decreasing solids fraction

6. only axial dependency, i.e. 1D model

7. solids are mono-sized, i.e. calculations are performed with a time dependent Sauter diameter growth

8. mean solids density for biomass-(product)-mixture

9. fixed total bed mass

10. no consideration of reactions in cyclones, standpipe, syphon and piping downstream of fluidized bed

5.1.1 Dense bottom zone

The gas is injected into the fluidized bed via a gas distributor. Furthermore, the lignin is fed pneumatically into the reactor via a vertical feeding lance. Numerous correlations for jet penetration have been derived in the past. A good summary of existing correlations is given in [284]. For the given pyrolysis parameters at the SPE pyrolysis plant the jet penetration depth of the fluidizing gas (650 °C, 4 bubble caps, 4.75 m/s) can be calculated by the correlations of [285] and [286] to be 6.2 cm and 6.0 cm, respectively. The feeding gas together with the pneumatic fed lignin penetrates (calculated by the same correlations at 650 °C, 6 mm lance diameter) between 7.8 and 8.4 cm into the bed. Thus, the bubble formation zone and jetting of feeding gas are neglected. The dense bottom zone consists of a solids free bubble and a suspension zone as depicted in Figure 5.2. The bubble phase has a volume fraction ε_b and the bubbles a height dependent rise velocity u_b, whereas the gas velocity in the suspension phase $u_{susp,g}$ is assumed to be constant. The bubble diameter d_b is also depending on dense zone height. If it reaches the reactor diameter d_R, after continuous growth at high superficial velocities u_0, it is fixed to the maximum of d_R. The correlations for calculation of the bed fluid dynamics have been derived in detail elsewhere [280, 281] and are summarized in Table 5.1.

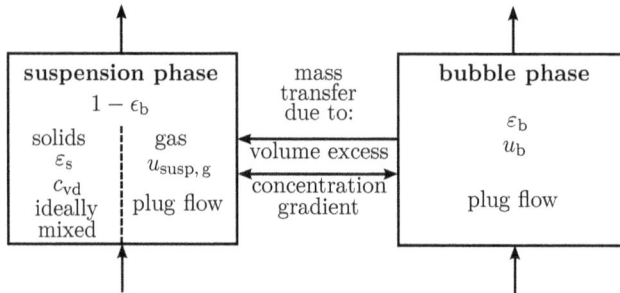

suspension phase	mass transfer due to:	bubble phase
$1 - \varepsilon_b$		ε_b
solids $\quad$ gas	volume excess	u_b
$\varepsilon_s \quad u_{susp,g}$	concentration	
c_{vd}	gradient	plug flow
ideally $\mid$ plug flow		
mixed		

Figure 5.2: Model structure of the dense bottom zone

Table 5.1: Correlations for calculation of fluid dynamics in the dense bottom zone

Not.	Quantity	Relation	Ref.	Eq.
Dimensionless numbers				
Ar	Archimedes number	$Ar = \frac{g \cdot \bar{d}_{sauter}^3}{\nu_f^2} \cdot \frac{(\bar{\rho}_s - \rho_f)}{\rho_f}$	[280]	(5.1)
Re_p	Reynolds number single particle	$Re_p = \frac{(u_0 - u_{mf}) \cdot \bar{d}_{sauter} \cdot \rho_f}{\eta_f}$	[280]	(5.2)
Re_{mf}	Reynolds number minimal fluidization	$Re_{mf} = \sqrt{25.7^2 + 0.0365 \cdot Ar} - 25.7$	[287]	(5.3)
Dense bed				
ε_b	bubble volume fraction	$\varepsilon_b = \frac{\dot{V}_b}{u_b}$	[280]	(5.4)
$\dot{V}_b$	visible volumetric bubble flow based on cross-sectional area of the bed in m/s	$\dot{V}_b = \zeta \cdot (u_0 - u_{mf})$	[280]	(5.5)

Table 5.1: (continued)

Not.	Quantity	Relation	Ref.	Eq.
ζ	parameter for $\dot{V}_b$	$\zeta = \begin{cases} 0.8 & \text{for Geldart A} \\ 1.45{\cdot}Ar^{-0.18} & \text{for Geldart B} \end{cases}$	[281] [280]	(5.6)
u_b	bubble rise velocity in m/s	$u_b = \dot{V}_b + 0.71\xi_b\sqrt{g{\cdot}d_b}$	[280]	(5.7)
ξ_b	parameter for u_b	$\xi_b = \begin{cases} 1.18 & d_R < 0.05\,\text{m} \\ 3.2{\cdot}d_R^{0.33} & 0.05\,\text{m} \leq d_R \leq 1\,\text{m for Geldart A} \\ 3.2 & d_R > 1\,\text{m} \end{cases}$	[281]	(5.8)
		$\xi_b = \begin{cases} 0.63 & d_R < 0.1\,\text{m} \\ 2\,d_R^{0.5} & 0.1\,\text{m} \leq d_R \leq 1\,\text{m for Geldart B} \\ 2 & d_R > 1\,\text{m} \end{cases}$	[280]	(5.9)
$d_{b,0}$	bubble diameter at distributor $(h = 0)$	$d_{b,0} = \begin{cases} 0.008{\cdot}\varepsilon_{b,0}^{\frac{1}{3}} & \text{for porous plate} \\ 1.3\left(\dot{V}_n^2{\cdot}\frac{1}{g}\right)^{0.2} & \begin{array}{l}\text{for technical}\\ \text{gas distributor,}\\ \dot{V}_n \stackrel{\wedge}{=} \text{volume flow through}\\ \text{single orifice}\end{array} \end{cases}$	[281] [280]	(5.10)
d_b	bubble diameter	$d_b = \begin{cases} \frac{dd_b}{dh} = \left(\frac{2\,\varepsilon_b(h)}{9\pi}\right)^{\frac{1}{3}} - \frac{d_b(h)}{3\lambda u_b(h)} & \text{for Geldart A} \\ \frac{dd_b}{dh} = \left(\frac{2\,\varepsilon_b(h)}{9\pi}\right)^{\frac{1}{3}} & \text{for Geldart B} \end{cases}$	[281] [280]	(5.11)
λ	mean duration of bubble life in s	$\lambda = 280{\cdot}\frac{u_{mf}}{g}$	[280]	(5.12)
c_{vd}	solids volume concentration in dense phase	$c_{vd} = (1-\varepsilon_{mf}){\cdot}\left(1-0.14\,Re_p^{0.4}{\cdot}Ar^{-0.13}\right)$	[280]	(5.13)
ε_s	solids volume fraction	$\varepsilon_s(h) = (1-\varepsilon_b){\cdot}c_{vd}$	[280]	(5.14)
$u_{b,g}$	gas velocity in bubble phase	$u_{b,g} = \frac{u_0 - u_{susp,g}{\cdot}(1-\varepsilon_b)}{\varepsilon_b}$		(5.15)
$u_{susp,g}$	gas velocity in suspension phase based on cross-sectional suspension area	$u_{susp,g} = \begin{cases} u_{mf} + \frac{1}{4}(u_0 - u_{mf}) & \begin{array}{l}\text{for technical}\\ \text{gas distributor}\end{array} \\ u_{mf} + \frac{1}{3}(u_0 - u_{mf}) & \text{for porous plate} \end{cases}$	[288] [289]	(5.16)

5.1.2 Mass transfer between phases

Due to gas concentration gradients between suspension and bubble phase, material transport has to be considered in the dense bottom zone. The transfer coefficient was defined for bubbling fluidized beds [290]. It holds

$$K_g = \frac{u_{mf}}{3} + \sqrt{\frac{4 \cdot D_{i,g} \cdot \varepsilon_{mf} \cdot u_b}{\pi d_b}} \quad . \tag{5.17}$$

As simplification binary diffusion is assumed. The binary diffusion coefficient $D_{i,g}$ of a component i in the gas g is calculated by the Fuller-method provided in [291].

$$D_{i,g}(\text{in cm}^2/\text{s}) = \frac{0.00143\text{cm}^2/\text{s} \cdot \left(\frac{T}{K}\right)^{1.75} \cdot \left[\left(\frac{M_i}{g/\text{mol}}\right)^{-1} + \left(\frac{M_g}{g/\text{mol}}\right)^{-1}\right]^{0.5}}{\frac{p}{\text{bar}} \cdot \sqrt{2} \cdot \left[\Sigma_{v,i}^{1/3} + \Sigma_{v,g}^{1/3}\right]^2} \tag{5.18}$$

The diffusion volume increments and the chemical structure used for calculation of the diffusion volume Σ_v of a component i are listed in Tables 5.2 and 5.3, respectively.

Table 5.2: Atomic diffusion volumes [291]

atomic and structural diffusion volume increments	
C	15.9
H	2.31
O	6.11
N_2	18.5

Table 5.3: Chemical structure of vapor phase components

chemical structure	
pyrolysis gas[†]	$C_{0.7}H_{1.1}O_{0.9}$
pyrolysis oil[†]	$C_{16.7}H_{19.5}O_{3.9}$
nitrogen	N_2

[†]from measured composition of pyrolysis gas and oil

The transfer between suspension and bubble phase occurs via their volume specific interface area a_b, which was defined by Werther and Hartge [282] as

$$a_b = \frac{6\,\varepsilon_b}{d_b} \quad . \tag{5.19}$$

The model assumes the gas in the suspension phase to have a constant velocity. Therefore, the excess gas created by reactions in the suspension phase induces a mass transfer from the suspension to the bubble phase. Excess volume can either evolve from heterogeneous reactions in which e.g. solid biomass reacts to form oil vapors and gas or from homogeneous reactions with an increase in moles (e.g. oil vapors react to form gases with smaller molar mass). Considering the molar balance and the ideal gas law the exchange rate K_q can be defined [292]. It is calculated from the sum of excess reaction rates of all gaseous components m in n reactions. In this work, the pressure gradient dp/dh is neglected.

$$K_q = -\frac{u_{mf}}{p} \cdot \overset{0}{\cancel{\frac{dp}{dh}}} + \frac{R \cdot T}{p} \cdot \sum_i^m \sum_j^n (r_{i,j} \cdot \varphi_{i,j}) \tag{5.20}$$

5.1.3 Dilute upper zone and entrainment

The correlations for the fluid dynamics in the dilute upper reactor zone are listed in Table 5.4. The dilute upper zone (cf. Figure 5.3) is a suspension with exponentially decreasing solids volume fraction. It decreases with increasing axial distance from the solids concentration at the inter-zone bed height $\varepsilon_s(h_{db})$ to a constant volume fraction above TDH $\varepsilon_{s,\infty}$. The decay constant was experimentally derived for particles with 240 µm [287]. The solids volume fraction above TDH is calculated from the equation of continuity and the solids entrainment flux above TDH. This entrained solids flux at TDH is calculated according to a correlation derived by Choi et al. [293]. The correlation has a broad range of validity as it was derived from experimental data from different units with various materials and operating conditions: $u_0 = 0.3$ to $7.0\,\mathrm{m/s}$, $d_R = 0.06$ to $1.0\,\mathrm{m}$ and $d_p = 0.05$ to $1.0\,\mathrm{mm}$ [294]. The net upwards solids velocity $u_{solid}(h)$ is assumed to have a reciprocal relationship to the solids volume fraction, as in the lower part of the dilute zone more particles move also in the downward direction. This velocity is the solids velocity used for the residence time calculation and thus to determine the heterogeneous plug flow reactions in this zone. It is assumed that the entrainment flux at the reactor outlet does depend on the solids volume fraction $\varepsilon_s(h = h_R)$ at h_R and the terminal velocity u_t.

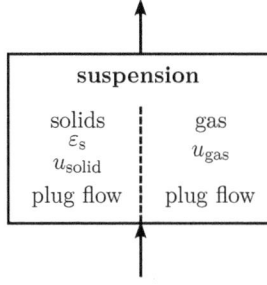

Figure 5.3: Model structure of the dilute upper zone

5.1.4 Solid properties

In pyrolysis processes the solids are usually formed in two ways (cf. Fig. 5.4). Either the pyrolyzed particles retain their particulate structure, examples are coal pyrolysis and combustion and pyrolysis of most lignocellulosic biomasses, or if a liquid is pyrolyzed this liquid forms a char layer on the bed material during pyrolysis (e.g. in the FCC process where the coke layer deactivates the FCC catalyst [295]). Latter is also possible if a liquid intermediate is formed due to melting of a solid feedstock and subsequent pyrolysis. The bed material is then granulated with a char layer.

As the biomass (virgin) is fed into the reactor, activated (active) and reacts to form char and volatiles (gas and oil) the mean solid properties of the sand-biomass-and-solid-

Table 5.4: Correlations for calculation of fluid dynamics in dilute upper zone

Not.	Quantity	Relation	Ref.	Eq.
ε_s	solids volume content	$\varepsilon_\mathrm{s}(h) = \varepsilon_{\mathrm{s},\infty} + [\varepsilon_\mathrm{s}(h_\mathrm{db}) - \varepsilon_{\mathrm{s},\infty}] \cdot e^{-a\cdot(h-h_\mathrm{db})}$	[280]	(5.21)
a	decay constant, in 1/m	$a = \frac{4/s}{u_0}$	[287]	(5.22)
$\varepsilon_{\mathrm{s},\infty}$	solids volume content above TDH	$\varepsilon_{\mathrm{s},\infty} = \frac{G_{\mathrm{s},\infty}}{\bar{\rho}_\mathrm{s}\cdot u_\mathrm{t}}$		(5.23)
u_t	single particle terminal velocity	$u_\mathrm{t} = U_\mathrm{t}^\star / [\rho_\mathrm{f}^2/(\eta_\mathrm{f}\cdot(\bar{\rho}_\mathrm{s}-\rho_\mathrm{f})\cdot g)]^{1/3}$	[287]	(5.24)
$U_\mathrm{t}^\star$	dimensionless velocity for u_t	$U_\mathrm{t}^\star = \left[18/D_\mathrm{p}^{\star2} + (2.335-1.744\cdot\psi_\mathrm{Wa})/\sqrt{D_\mathrm{p}^\star}\right]^{-1}$	[287]	(5.25)
$D_\mathrm{p}^\star$	dimensionless diameter for u_t	$D_\mathrm{p}^\star = \bar{d}_\mathrm{sauter}\cdot\left[(\rho_\mathrm{f}\cdot(\bar{\rho}_\mathrm{s}-\rho_\mathrm{f})\cdot g)/\eta_\mathrm{f}^2\right]^{1/3}$	[287]	(5.26)
$G_{\mathrm{s},\infty}$	entrainment rate above TDH in kg/(m² · s)	$G_{\mathrm{s},\infty} = Ar^{0.5}\cdot\exp\left[6.92-2.11F_\mathrm{g}^{0.303}-\frac{13.1}{F_\mathrm{d}^{0.902}}\right]\cdot\frac{\eta_\mathrm{f}}{\bar{d}_\mathrm{sauter}}$	[293]	(5.27)
F_g	gravity minus buoyancy force per projection area of particle in Pa	$F_\mathrm{g} = g\cdot\bar{d}_\mathrm{sauter}\cdot(\bar{\rho}_\mathrm{s}-\rho_\mathrm{f})$	[293]	(5.28)
F_d	drag force on the particle per projection area of particle in Pa	$F_\mathrm{d} = C_\mathrm{d}\cdot\rho_\mathrm{f}\cdot\frac{u_0^2}{2}$	[293]	(5.29)
C_d	drag coefficient on the particle surface based on superficial gas velocity, dimensionless	$C_\mathrm{d} = \begin{cases} 24/Re_\mathrm{p} & \text{for } Re_\mathrm{p}(u_0) \leq 5.8 \\ 10/Re_\mathrm{p}^{0.5} & \text{for } 5.8 < Re_\mathrm{p}(u_0) \leq 540 \\ 0.43 & \text{for } 540 < Re_\mathrm{p}(u_0) \end{cases}$	[293]	(5.30)
$u_\mathrm{solid}(h)$	mean solid velocity in dilute zone in m/s	$u_\mathrm{solid} = \frac{G_{\mathrm{s},\infty}}{\bar{\rho}_\mathrm{s}\cdot\varepsilon_\mathrm{s}(h)}$		(5.31)
G_s	entrainment rate at reactor outlet in kg/(m² · s)	$G_\mathrm{s} = \varepsilon_\mathrm{s}(h=h_\mathrm{R})\cdot\bar{\rho}_\mathrm{s}\cdot u_\mathrm{t}$		(5.32)

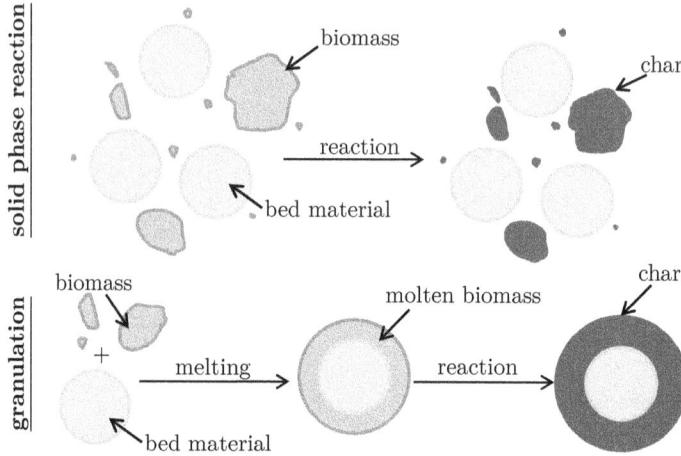

Figure 5.4: Solids formation in pyrolysis processes

product-mixture in the bed change. Both mean particle diameter and mean density have an influence on the fluid dynamics. In case of surface proportional granulation, the particle growth rate [296] is defined as

$$\frac{\mathrm{d}d}{\mathrm{d}t} = \frac{2 \cdot \dot{m}}{\rho \cdot A_{\text{total}}} \qquad . \tag{5.33}$$

Solving Eq. 5.33 gives the mean Sauter diameter. The solution for char only is

$$\bar{d}_{sauter} = d_{\text{QS}} \cdot \sqrt[3]{\frac{\dot{m}_{\text{char}}}{G_{\text{s}} \cdot A_{\text{R}} - \dot{m}_{\text{char}}} \cdot \frac{\rho_{\text{QS}}}{\rho_{\text{char}}} + 1} \qquad . \tag{5.34}$$

In case of incomplete conversion the total mean diameter, resulting from deposition of active and char, is calculated from sequential layering. For biomasses, that do not melt to form sand-char granulates the change in particle size is not considered. For all biomasses the mean density is calculated from the volume fractions x of each solid component i in the solid mixture.

$$\bar{\rho}_{\text{s}} = \sum_i \rho_i \cdot x_i \tag{5.35}$$

5.1.5 Change of superficial velocity due to heterogeneous reactions

The change of moles due to the reaction, i.e. evolution of gases and vapors released from the solid lignin matrix, directly increases the gas velocities in the riser. The differential equation for the change of superficial gas velocity with reactor height was in detail derived by Sitzmann et al. [297] and is described in short here. With the reaction kinetics described below, the total change in molar flow for all involved reactions in a discrete volume element $\mathrm{d}V = A_{\text{R}} \cdot \mathrm{d}h$ can be calculated

$$\mathrm{d}n = A_{\text{R}} \cdot \mathrm{d}h \cdot \sum_i r_i \cdot \varphi_i(h) \qquad . \tag{5.36}$$

Applying the ideal gas law the change in molar flow can be described by

$$\frac{d\dot{n}}{dh} = \frac{1}{R \cdot T} \cdot \left[\left(\frac{dp}{dh} \right) \cdot \dot{V} + \left(\frac{d\dot{V}}{dh} \right) \cdot p \right] \quad . \tag{5.37}$$

Substituting $\dot{V}$ by $A_R \cdot u$ and solving for u yields the differential relationship for change in axial superficial gas velocity

$$\frac{du}{dh} = \frac{1}{p} \cdot \left[-u \cdot \cancelto{0}{\left(\frac{dp}{dh} \right)} + \frac{R \cdot T}{A_R} \cdot \left(\frac{d\dot{n}}{dh} \right) \right] \quad . \tag{5.38}$$

In this work the pressure gradient dp/dh is neglected.

5.1.6 Residence times

The average residence time of both solids τ_s and gas τ_g in the riser reactor are defined as follows

$$\tau_s = \tau_{s,db} + \tau_{s,fb} = \frac{m_{db}}{G_s \cdot A_R} + \sum \frac{\Delta h}{u_{solid}(h)} \tag{5.39}$$

and

$$\tau_g = \sum \frac{\Delta h}{u(h)/(1 - \varepsilon_s(h))} \quad . \tag{5.40}$$

5.2 Reaction model and mass balances

The reaction scheme of lignocellulosic biomass pyrolysis is described by a micro-particle model of Miller and Bellan [221] (Figure 5.5). It is composed of a superimposed reaction scheme for the three main lignocellulosic biomass components j (cellulose, hemicellulose, and lignin). Each biomass component (virgin) is activated in a preliminary reaction step (reaction 1). The activated intermediate (active) then reacts in parallel primary reactions 2 and 3 to pyrolysis oil and char + pyrolysis gas (gas I), respectively. The latter two are formed in reaction 3 in a mass ratio of $\hat{X}$ and $(\hat{X} - 1)$, respectively.

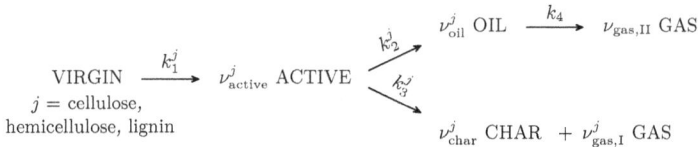

$$\text{VIRGIN} \xrightarrow{k_1^j} \nu_{active}^j \text{ ACTIVE} \begin{array}{c} \xrightarrow{k_2^j} \nu_{oil}^j \text{ OIL} \xrightarrow{k_4} \nu_{gas,II} \text{ GAS} \\ \\ \xrightarrow{k_3^j} \nu_{char}^j \text{ CHAR} + \nu_{gas,I}^j \text{ GAS} \end{array}$$

j = cellulose, hemicellulose, lignin

Figure 5.5: Reaction scheme for pyrolysis of lignocellulose micro particles according to Miller and Bellan [221]

The char formation mass ratio $\hat{X}$ differs for each biomass component but is assumed to be temperature independent [221]. In a secondary reaction (reaction 4) pyrolysis oil reacts further to form pyrolysis gas (gas II). Miller and Bellan [221] assume the reaction rate constant for the secondary reaction k_4 to be independent of the biomass component, while all other reaction rate constants are different for the three biomass components j. The temperature dependency of the reaction rate constants k_i^j is given by the Arrhenius equation

$$k_i^j = k_{0,i}^j \cdot e^{-\frac{E_{A,i}^j}{R \cdot T}} \quad . \tag{5.41}$$

The kinetic model parameters are listed in Table 5.5. For stoichiometry the molar masses of the pseudo-components are defined. The molar mass of lignin, pyrolysis oil and pyrolysis gas is approximated by the composition determined in this work. The molar masses of cellulose and hemicellulose are calculated from a degree of polymerization of 3000 and 200, respectively and the molar mass of repeating unit $C_6H_{10}O_5$ (cf. Section 2.1.1). For the activated intermediates it is assumed that no change in molar mass occurs. The molar mass of char is equal to carbon. The molar masses are given together with the mass ratios of char formation $\hat{X}^j$ in Table 5.6.

Table 5.5: Parameters for reaction kinetics, [221]

reaction rate constant	component					
	cellulose		hemicellulose		lignin	
	k_0, 1/s	E_A, kJ/mol	k_0, 1/s	E_A, kJ/mol	k_0, 1/s	E_A, kJ/mol
k_1	$2.80 \cdot 10^{19}$	242.4	$2.10 \cdot 10^{16}$	186.7	$9.60 \cdot 10^{8}$	107.6
k_2	$3.28 \cdot 10^{14}$	196.5	$8.75 \cdot 10^{15}$	202.4	$1.50 \cdot 10^{9}$	143.8
k_3	$1.30 \cdot 10^{10}$	150.5	$2.60 \cdot 10^{11}$	145.7	$7.70 \cdot 10^{6}$	111.4
k_4	$4.28 \cdot 10^{6}$	108	$4.28 \cdot 10^{6}$	108	$4.28 \cdot 10^{6}$	108

Table 5.6: Molar mass M and char formation mass ratio $\hat{X}$ for reaction 3 of reaction scheme components

component	M, kg/kmol	$\hat{X}^{\dagger}$, kg/kg
cellulose	$3000 \cdot 162$	0.35
hemicellulose	$200 \cdot 162$	0.60
lignin	3500	0.75
char	12.01	-
gas	24.6	-
oil	280	-

† from [221]

The stoichiometric factors ν_i^j are defined as follows:

$$\nu_{\text{virgin}}^j = \nu_{\text{active}}^j = 1 \quad ,$$
$$\nu_i^j = \frac{M_j}{M_i} \qquad \text{for } i = \text{oil, gas II} \quad ,$$
$$\nu_{\text{gas I}}^j = \frac{M_j \cdot (1 - \hat{X}^j)}{M_{\text{gas}}} \qquad \text{and}$$
$$\nu_{\text{char}}^j = \frac{M_j \cdot \hat{X}^j}{M_{\text{char}}} \quad .$$

5.2.1 Material balances

As the bubble phase is assumed to be solids free the solid reactions take place in the suspension phases in the dense bottom and dilute upper zone only. Whereas the secondary reaction from oil to gas takes place also in the bubble phase. For all phases and for both gas and solids the overall mole balance holds. For the stationary model, it can be written as

$$\cancel{\frac{\mathrm{d}n_i}{\mathrm{d}t}}^{0} = 0 = \Sigma \dot{n}_i^{\text{in}} - \Sigma \dot{n}_i^{\text{out}} \pm \Sigma \dot{n}_i^{\text{reac}} \quad . \tag{5.42}$$

All molar balances are calculated for the three main biomass components cellulose, hemicellulose, and lignin separately, based on the fed biomass composition of these components, denoted with j. Thus, giving three sets of equations for the biomass components plus a set for the cumulative molar flow and concentration profile for the reaction products of the overall biomass.

5.2.1.1 Dense bottom zone

Solid phase balances

For the dense bottom zone's suspension phase the solid reaction volume is assumed to be ideally mixed. As the reactor is divided into discrete volume elements, continuous flow stirred-tank reactor (CSTR) elements are used in parallel. The fed biomass is assumed to be distributed to the discrete volume elements proportionately to the fraction of solids mass in that element. The mole balance for component i is

$$0 = \dot{n}_i^{\text{in}} \cdot \frac{m_{\text{s}}(\Delta h)}{m_{\text{db}}} - \dot{V}^{\text{out}} \cdot c_i + V_{\text{R}} \cdot r_i \tag{5.43}$$

with the volume flow out of the volume element, which is defined by the phase's reaction volume V_{R} and the residence time $\tau_{\text{s, db}}$ in the volume element

$$\dot{V}^{\text{out}} = \frac{V_{\text{R}}}{\tau_{\text{s, db}}} \quad . \tag{5.44}$$

The residence time, due to perfect mixing and mono-sized volume elements, is assumed to be the same in all volume elements and equal to

$$\tau_{\text{s, db}} = \frac{m_{\text{db}}}{G_{\text{s}} \cdot A_{\text{R}}} \quad . \tag{5.45}$$

The reaction volume of the solid phase is depending on the volume fraction of the suspension phase $1 - \varepsilon_b$ and the volume fraction of solids in the suspension c_{vd}. It holds

$$V_R = A_R \cdot (1 - \varepsilon_b(h)) \cdot c_{vd} \cdot \Delta h \qquad . \tag{5.46}$$

Equation 5.46 can be rewritten with the volume of the volume element $V_{element} = A_R \cdot \Delta h$ to obtain the volume fraction φ_i of the reacting phase

$$\varphi_i = \frac{V_R}{V_{element}} = \frac{V_R}{A_R \cdot \Delta h} = (1 - \varepsilon_b(h)) \cdot c_{vd} \qquad . \tag{5.47}$$

Three reactants i in the pyrolysis scheme (cf. Figure 5.5) are solid. Thus, the reaction rates are given for virgin, active and char by:

$$r_{virgin} = -k_1 \cdot c_{virgin} \qquad ,$$
$$r_{active} = k_1 \cdot c_{virgin} - k_2 \cdot c_{active} - k_3 \cdot c_{active} \qquad \text{and}$$
$$r_{char} = \frac{\nu_{char}}{\nu_{active}} \cdot k_3 \cdot c_{active} \qquad .$$

Gas phase balances

The gas passes all phases in plug flow. The change in molar flow from the reactions and the net convective transport through the volume element are considered. Furthermore, gas is exchanged between the bubble and suspension phase via diffusive and convective mass transport. The volume excess due to the change in molar flow, which results from the reactions is transferred to the bubble phase. Thus, the overall mole balance gives

$$0 = -\frac{d\dot{n}_i^{conv}}{dh} \cdot dh \pm \Sigma \dot{n}_i^{reac} \pm \dot{n}_i^{excess} \pm \dot{n}_i^{transfer} \qquad . \tag{5.48}$$

Suspension phase
For the suspension phase the terms in Eq. 5.48 are given by:

$$\dot{n}_i^{conv} = u_{susp}(h) \cdot A_R \cdot (1 - \varepsilon_b(h)) \cdot (1 - c_{vd}) \cdot c_{i,susp}(h)$$
$$\dot{n}_i^{reac} = r_i \cdot \varphi_i(h) \cdot A_R \cdot dh$$
$$\dot{n}_i^{excess} = -K_q(h) \cdot c_{i,susp}(h) \cdot A_R \cdot (1 - \varepsilon_b(h)) \cdot (1 - c_{vd}) \cdot dh$$
$$\dot{n}_i^{transfer} = -K_g(h) \cdot a_b(h) \cdot A_R \cdot (c_{i,susp} - c_{i,b}) \cdot dh \qquad .$$

The final balance after canceling $V_{element} = A_R \cdot \Delta h$ is

$$0 = -(1 - c_{vd}) \cdot \frac{d(u_{susp}(h) \cdot (1 - \varepsilon_b(h)) \cdot c_{i,susp}(h))}{dh} + r_i \cdot \varphi_i(h) \tag{5.49}$$
$$- K_q(h) \cdot c_{i,susp}(h) \cdot (1 - \varepsilon_b(h)) \cdot (1 - c_{vd}) - K_g(h) \cdot a_b(h) \cdot (c_{i,susp} - c_{i,b}) \qquad .$$

The reaction terms $r_i \cdot \varphi_i(h)$ are valid for gas, oil and inert (gas). It holds:

$$r_{\text{oil}} \cdot \varphi_{\text{oil}}(h) = \frac{\nu_{\text{oil}}}{\nu_{\text{active}}} \cdot k_2 \cdot c_{\text{active, susp}} \cdot (1 - \varepsilon_{\text{b}}(h)) \cdot c_{\text{vd}}$$
$$- k_4 \cdot c_{\text{oil, susp}} \cdot (1 - \varepsilon_{\text{b}}(h)) \cdot (1 - c_{\text{vd}})$$

$$r_{\text{gas}} \cdot \varphi_{\text{gas}}(h) = \frac{\nu_{\text{gas, I}}}{\nu_{\text{active}}} \cdot k_3 \cdot c_{\text{active, susp}} \cdot (1 - \varepsilon_{\text{b}}(h)) \cdot c_{\text{vd}}$$
$$+ \frac{\nu_{\text{gas, II}}}{\nu_{\text{oil}}} \cdot k_4 \cdot c_{\text{oil, susp}} \cdot (1 - \varepsilon_{\text{b}}(h)) \cdot (1 - c_{\text{vd}})$$

$$r_{\text{inert}} \cdot \varphi_{\text{inert}}(h) = 0 \quad .$$

Bubble phase

Analogously to the suspension phase, the terms in the bubble phase are defined as below. The excess molar flow $\dot{n}_i^{\text{excess}}$ and the material transfer molar flow $\dot{n}_i^{\text{transfer}}$ have opposite sign compared to the same terms in the dense zone's suspension phase as the transfer occurs between these phases.

$$\dot{n}_i^{\text{conv}} = u_{\text{b}}(h) \cdot A_{\text{R}} \cdot \varepsilon_{\text{b}}(h) \cdot c_{i, \text{b}}(h)$$
$$\dot{n}_i^{\text{reac}} = r_i \cdot \varphi_i(h) \cdot A_{\text{R}} \cdot \text{d}h$$
$$\dot{n}_i^{\text{excess}} = K_{\text{q}}(h) \cdot c_{i, \text{susp}}(h) \cdot A_{\text{R}} \cdot (1 - \varepsilon_{\text{b}}(h)) \cdot (1 - c_{\text{vd}}) \cdot \text{d}h$$
$$\dot{n}_i^{\text{transfer}} = K_{\text{g}}(h) \cdot a_{\text{b}}(h) \cdot A_{\text{R}} \cdot (c_{i, \text{susp}} - c_{i, \text{b}}) \cdot \text{d}h$$

The final balance can be derived the same way as the suspension phase balance

$$0 = - \frac{\text{d}(u_{\text{b}}(h) \cdot \varepsilon_{\text{b}}(h) \cdot c_{i, \text{b}}(h))}{\text{d}h} + r_i \cdot \varphi_i(h) \tag{5.50}$$
$$+ K_{\text{q}}(h) \cdot c_{i, \text{susp}}(h) \cdot (1 - \varepsilon_{\text{b}}(h)) \cdot (1 - c_{\text{vd}}) + K_{\text{g}}(h) \cdot a_{\text{b}}(h) \cdot (c_{i, \text{susp}} - c_{i, \text{b}}) \quad .$$

The reaction terms $r_i \cdot \varphi_i(h)$ are valid for gas, oil and inert (gas). It holds:

$$r_{\text{oil}} \cdot \varphi_{\text{oil}}(h) = -k_4 \cdot c_{\text{oil, b}} \cdot \varepsilon_{\text{b}}(h)$$
$$r_{\text{gas}} \cdot \varphi_{\text{gas}}(h) = \frac{\nu_{\text{gas, II}}}{\nu_{\text{oil}}} \cdot k_4 \cdot c_{\text{oil, b}} \cdot \varepsilon_{\text{b}}(h)$$
$$r_{\text{inert}} \cdot \varphi_{\text{inert}}(h) = 0 \quad .$$

5.2.1.2 Dilute upper zone

For the upper dilute zone, plug flow is assumed for both solid and gas phase.

Solid phase balances

For the solid phase in the upper dilute zone reactions for virgin, active and char are considered. As the dilute zone solely consists of a suspension phase, the molar balance can be reduced to two terms:

$$\dot{n}_i^{\text{conv}} = u_{\text{s}}(h) \cdot A_{\text{R}} \cdot \varepsilon_{\text{s}}(h) \cdot c_i(h)$$
$$\dot{n}_i^{\text{reac}} = r_i \cdot \varphi_i(h) \cdot A_{\text{R}} \cdot \text{d}h \quad .$$

The final balance for every volume element is

$$0 = -\frac{d(u_s(h) \cdot \varepsilon_s(h) \cdot c_i(h))}{dh} + r_i \cdot \varphi_i(h) \qquad . \tag{5.51}$$

It holds further:

$$r_{\text{virgin}} \cdot \varphi_{\text{virgin}}(h) = -k_1 \cdot c_{\text{virgin}} \cdot \varepsilon_s(h)$$
$$r_{\text{active}} \cdot \varphi_{\text{active}}(h) = k_1 \cdot c_{\text{virgin}} \cdot \varepsilon_s(h) - k_2 \cdot c_{\text{active}} \cdot \varepsilon_s(h) - k_3 \cdot c_{\text{active}} \cdot \varepsilon_s(h)$$
$$r_{\text{char}} \cdot \varphi_{\text{char}}(h) = \frac{\nu_{\text{char}}}{\nu_{\text{active}}} \cdot k_3 \cdot c_{\text{active}} \cdot \varepsilon_s(h) \qquad .$$

Gas phase balances

In the suspension gas phase of the dilute upper zone the convective and reaction terms are:

$$\dot{n}_i^{\text{conv}} = u(h) \cdot A_R \cdot (1 - \varepsilon_s(h)) \cdot c_i(h)$$
$$\dot{n}_i^{\text{reac}} = r_i \cdot \varphi_i(h) \cdot A_R \cdot dh \qquad .$$

The final balance for every gaseous component is

$$0 = -\frac{d(u(h) \cdot (1 - \varepsilon_s(h)) \cdot c_i(h))}{dh} + r_i \cdot \varphi_i(h) \qquad . \tag{5.52}$$

Oil and gas are produced in the heterogeneous reactions from active and oil cracked in a secondary reaction to form gas II. The inert fluidizing gas does not partake in the reaction.

$$r_{\text{oil}} \cdot \varphi_{\text{oil}}(h) = \frac{\nu_{\text{oil}}}{\nu_{\text{active}}} \cdot k_2 \cdot c_{\text{active}} \cdot \varepsilon_s(h) - k_4 \cdot c_{\text{oil}} \cdot (1 - \varepsilon_s(h))$$
$$r_{\text{gas}} \cdot \varphi_{\text{gas}}(h) = \frac{\nu_{\text{gas,I}}}{\nu_{\text{active}}} \cdot k_3 \cdot c_{\text{active}} \cdot \varepsilon_s(h) + \frac{\nu_{\text{gas,II}}}{\nu_{\text{oil}}} \cdot k_4 \cdot c_{\text{oil}} \cdot (1 - \varepsilon_s(h))$$
$$r_{\text{inert}} \cdot \varphi_{\text{inert}}(h) = 0$$

Yields and conversion

The product yields of gas, oil, and char are defined as the proportionate sum of the yields for each of the three biomass components j = cellulose, hemicellulose, and lignin. Which themselves are calculated by the quotient of produced mass and the feed of virgin biomass component j.

$$Y_{\text{oil}} \quad = \quad \sum_j w_j \cdot Y_{\text{oil},j} \quad = \quad \sum_j w_j \cdot \frac{\dot{n}_{\text{oil},j}}{\dot{n}_{\text{virgin},j}} \cdot \frac{M_{\text{oil}}}{M_j} \tag{5.53}$$

$$Y_{\text{gas}} \quad = \quad \sum_j w_j \cdot Y_{\text{gas},j} \quad = \quad \sum_j w_j \cdot \frac{\dot{n}_{\text{gas},j}}{\dot{n}_{\text{virgin},j}} \cdot \frac{M_{\text{gas}}}{M_j} \tag{5.54}$$

$$Y_{\text{char}} \quad = \quad \sum_j w_j \cdot Y_{\text{char},j} \quad = \quad \sum_j w_j \cdot \frac{\dot{n}_{\text{char},j}}{\dot{n}_{\text{virgin},j}} \cdot \frac{M_{\text{gas}}}{M_j} \tag{5.55}$$

The conversion is defined as the proportionate sum of ratios of converted biomass component j and the total feed of virgin biomass component j as follows

$$X = \sum_j w_j \cdot \left[\frac{\dot{n}_{\text{virgin},j} - (\dot{n}_{\text{virgin},\text{out},j} + \dot{n}_{\text{active},\text{out},j})}{\dot{n}_{\text{virgin},j}} \right] \tag{5.56}$$

with $\dot{n}_{\text{virgin},\text{out},j}$ and $\dot{n}_{\text{active},\text{out},j}$ the unconverted mass flows of virgin and active leaving the reactor.

5.3 Literature data used for model validation

5.3.1 CFB air-blown pyrolysis of pine wood, CERTH, Greece

At the Chemical Process and Energy Resources Institute (CPERI), Centre for Research & Technology Hellas (CERTH) a circulating fluidized bed air-blown pyrolyzer was developed in 2007. First, the plant was studied at cold conditions, i.e. ambient temperature, to investigate the fluid dynamical behavior, such as the recirculation rates [88]. With the obtained design, the pyrolysis of pine wood chips was studied in the system that integrates a char combustor to provide autothermal operation of the pyrolysis reactor [89]. The char combustion takes place in a stationary fluidized bed, which is fluidized with air. The cross-sectional area of the reactor above the combustion zone reduces to a final riser diameter of 50 mm. Thus, the particles heated by char combustion are directly transferred into the riser (with 2 m height) and enable the endothermic pyrolysis reaction. The mass flow of 10 kg pine wood is fed by a screw feeder into the riser. The particles in the riser that are entrained together with the product gases are separated from the latter in two cyclones and fed back into the combustion part of the integrated system. Sand is used as a heat carrier with a particle size that is not specified in the publications [88, 89, 298]. The known plant dimensions and operational parameters are summarized in Figure 5.6 and Table 5.7.

In total five experiments are listed in the publication [89]. The yields of pyrolysis oil, gas, and char are not explicitly listed, but the mass flows of liquid and gaseous product and gas composition are given. Due to the integrated combustion-pyrolysis process, the gas contains both combustion and pyrolysis gases. The oil yield is calculated as the mass fraction of the sum of liquid product mass flows over the dry biomass mass flow. The char yield is approximated by the consumed oxygen mass flow and a CO/CO_2 ratio of 0.1 and the biomass mass flow. In permanent gas oxidation the rest of the oxygen is consumed. Table 5.8 lists the pyrolysis temperature, gas residence time and calculated yields.

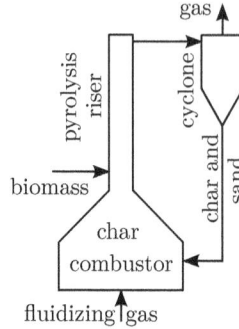

Figure 5.6: Air-blown pyrolysis reactor at CERTH, simplified from [88]

The parameters listed in Table 5.7 are used for simulation. The initial superficial gas velocity entering the riser reactor of the autothermal system is calculated from the sum of flue gas mass fluxes and the riser temperature. The heat carrier sand is assumed to have a mean particle size of 200 μm, which is the same as used for the simulation of this works lignin pyrolysis. Furthermore, the holdup of bed material in the riser is assumed to be 0.17 kg, as due to the specific plant design the formation of a dense bottom zone in the riser is not to be expected.

Table 5.7: Plant dimensions and operational parameters for CFB pyrolysis of pine wood chips [88, 89]

parameter	value
riser height	2 m
riser diameter	0.05 m
biomass	pine wood chips
composition[†] [143]	50 wt.-% C | 27 wt.-% H | 23 wt.-% L
feed mass flow	8.44 to 10.45 kg/h
bed material	sand
vapor residence time	0.37 to 0.46 s
superficial gas velocity	3.85 to 5.31 m/s
pyrolysis temperature	496 to 650 °C

[†]C: cellulose, H: hemicellulose, and L: lignin

Table 5.8: Experimental conditions and results for CFB pyrolysis of pine wood chips

exp.	ϑ	τ_g	yield, wt.-%			$\dot{m}_b$	d_p
	°C	s	gas	oil	char	kg/s	mm
R7	581	0.374	44.2	48.3	7.5	8.44	1.5 - 2
R8	605	0.388	29.1	64.7	6.3	10.45	1 - 1.5
R9	579	0.396	27.9	66.4	5.7	10.2	1 - 1.5
R10	496	0.458	19.0	74.9	6.1	8.96	1 - 1.5
R11	550	0.373	29.2	64.7	6.1	10.22	1.5 - 2

5.3.2 Bubbling fluidized bed pyrolysis of biomass, University of Waterloo, Canada

A pilot plant for bubbling fluidized bed pyrolysis was developed at the Department of Chemical Engineering, University of Waterloo, Waterloo, Ontario. Both plant description and experimental results for various biomass have been reported [51, 83, 84]. Reactor dimensions and operational parameters are given in Figure 5.7 and Table 5.9. Bed fluidization is achieved by introduction of recycled product gas. The biomass particles with a diameter of about 600 μm are dosed by a screw feeder into a pneumatical feeding system. Injection of the particles is carried out from the top, slightly below bed surface. The particle size of the inert bed material is selected in such a way that the produced char is entrained with the product gas stream, while the inert sand remains in the bed; but the bed material particle size is not explicitly specified in the publication. A cyclone separates the char from the product vapors, which are separated into condensed liquid and permanent gas by a condenser. The vapor phase residence time is defined as the ratio of reactor volume and inlet gas volume flow [84]. A selection of the published yields is given in Table 5.10.

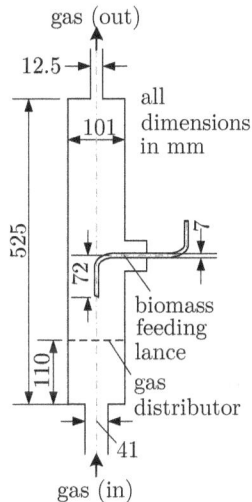

Figure 5.7: BFB pyrolysis reactor, University of Waterloo, Canada, simplified from [84]

As mentioned above, the bed material particle size is selected in such a way that char particles are entrained, while the sand is retained. Therefore, in the model, a particle size of 300 μm is selected to mimic this behavior. Furthermore, the solids volume fraction at minimal fluidization is assumed to be 0.5. The solids holdup in the bed is set to be 1.0 kg. In the model, the resulting bed height is about 0.2 m (about half reactor height). With the selected input parameters the solids entrainment is too high. Therefore, the decay constant is adjusted to satisfy the correlation $a \cdot u_0 = 25$. Calculation with these assumptions gives a solids residence time in the magnitude of 20 s, which is reasonable for the experimental setup, and results in nearly complete conversion. In the appendix, Table A.2 the input parameters are listed.

Table 5.9: Plant dimensions and operational parameters for bubbling fluidized bed pyrolysis of biomass [51, 84]

parameter	value
riser height	0.415 m
riser diameter	0.101 m
biomass	maple wood, aspen wood, wheat straw
composition[†], wt.-%	maple wood[‡]: 40.0 C, 38.0 H, 22.0 L
	aspen wood[*]: 50.2 C, 31.6 H, 18.2 L
	wheat straw[*]: 38.2 C, 38.4 H, 23.4 L
feed mass flow	1.5 to 3 kg/h
bed material	sand
vapor residence time	0.5 to 0.75 s
superficial gas velocity	0.54 to 0.81 m/s
pyrolysis temperature	425 to 625 °C

[†]C: cellulose, H: hemicellulose and L: lignin, [‡][221], [*][299], [*][27]

Table 5.10: Experimental conditions and results for bubbling fluidized bed pyrolysis of various biomass [84]

biomass	ϑ	τ_g	yield, wt.-%			$\dot{m}_b$
	°C	s	gas	oil	char	kg/s
maple wood	480	0.5	10.51	73.44	12.41	2.169
maple wood	500	0.5	12.31	73.06	9.24	2.573
maple wood	530	0.5	16.51	69.41	10.32	2.594
aspen wood	425	0.616	5.95	59.68	30.51	2.391
aspen wood	465	0.584	8.53	72.67	18.88	1.709
aspen wood	500	0.55	12.45	75.06	12.15	2.238
aspen wood	500	0.55	12.07	77.75	11.2	1.97
aspen wood	540	0.539	21.22	71.05	8.99	1.281
aspen wood	625	0.52	36.65	44.37	7.81	1.001
wheat straw	500	0.75	18.92	18.92	23.66	1.919
wheat straw	500	0.6	13.61	13.61	20.6	1.717
wheat straw	520	0.54	15.67	15.67	16.85	1.114
wheat straw	525	0.57	8.63	8.63	24.17	2.762
wheat straw	550	0.71	23.88	23.88	19.13	2
wheat straw	575	0.52	23.24	23.24	17.3	1.587

5.4 Flowsheet modeling of an integrated pyrolysis-combustion process

5.4.1 Flowsheet model components

For assessment of energetic process performance of an integrated pyrolysis–combustion process and to demonstrate the practicability of the derived pyrolysis model for biorefinery flowsheet simulations further unit models are implemented in ACM® (cf. Figure 5.8). Firstly, the fluidized bed pyrolysis model is extended with an overall energy balance. The heat of pyrolysis reaction is taken from a fit to the experimentally determined energy requirement [92] (cf. Fig. A.2a). Secondly, a fluidized bed combustion model for pyrolysis char and a gas burner are implemented. The fluidized bed combustor model is described below. For both by-product combustion reactors the mass and energy balances are calculated. For simplification, the empirical fit correlations for combustion heat are calculated from the experimental higher heating values of gas and char (cf. Figure A.1). Thirdly, ideal models for heat exchange, gas-liquid separation (liquid condensation), gas-solid separation, and solid mixing are included. Ideal separation is assumed, reactions and heat losses for the ideal models are neglected, while mass and energy conservation are considered for each component. In a stream the total enthalpy flow is defined as the sum of enthalpy flows of the single stream components i: $\sum_i \dot{H}_i = \sum \dot{m}_i \cdot c_{p,i} \cdot \Delta \vartheta$ (stream to ambient).

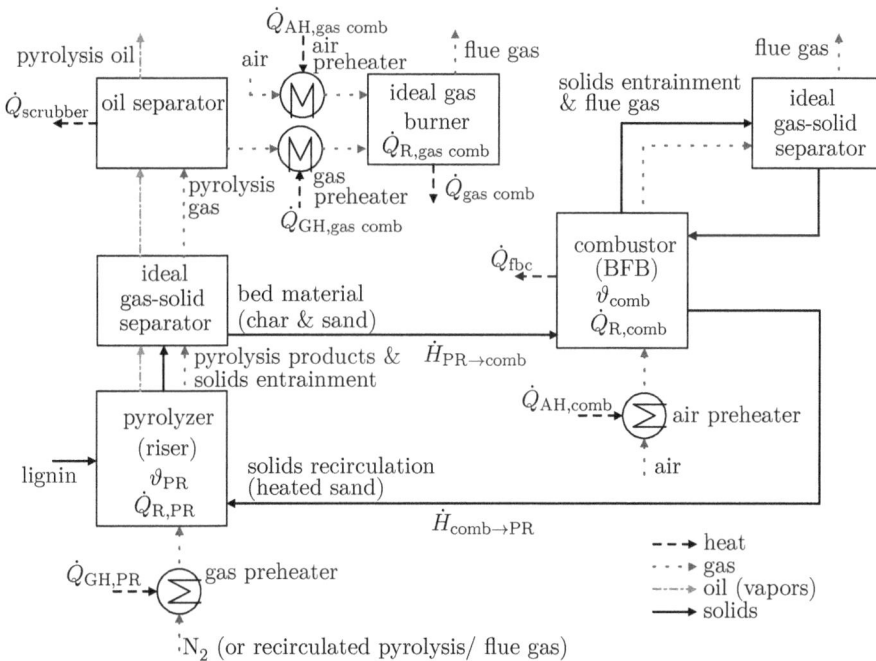

Figure 5.8: Flowsheet of integrated pyrolysis-combustion process

Fluidized bed char combustor

The fluidized bed model, used for the pyrolysis reactor, was originally developed for fluidized bed combustion processes. With minor adaptions, it was applied to conventional fluidized bed combustion [289, 300] as well as modeling of the fuel reactor of a chemical looping system with solid fuel feed [279]. Thus, the fluidized bed combustion model, applied in this work, is mainly based on these previous works. Also, the fluidized bed combustion reactor consists of a dense bottom zone and a dilute upper zone. The fluid dynamics are the same as in the pyrolysis reactor. In contrast the mass balances are implemented for the main char combustion reactions: first the heterogeneous char oxidation to CO and CO_2 and second the homogeneous oxidation of CO to CO_2 (cf. Section A.3). For simplification, other char constituents like e.g. sulfur, oxygen or hydrogen are neglected. Neglection sulfur, oxygen, and hydrogen is especially valid for char of large carbonization degree produced at high pyrolysis temperature. For flowsheet simulation, the overall mass balance of the reactor includes a drain off from dense bed (mass balance according to Puettmann et al. [278]). The single particle shrinkage, due to combustion of the char layer, is represented by the Shrinking Particle Inert Core (SPIC) model (considering the inert sand core) [92].

5.4.2 Flowsheet model of SPE plant with integrated char combustion

The setup of the flowsheet simulated in ACM® is depicted in Fig. 5.8. It is a representation of the experimental SPE pyrolysis setup extended with a model for by-product combustion. Thereby, the energy demand of endothermic pyrolysis can – depending on operating conditions – be supplied by combustion of pyrolysis gas and char. The pyrolyzer is coupled with a bubbling fluidized bed char combustion reactor. The heat is transferred to the pyrolyzer by the difference of enthalpy flow of the circulating bed material. The char laden bed material exiting the pyrolyzer flows via an ideal gas-solid separator into the fluidized bed combustor, where the char is combusted and the sand heated. To complete the bed material mass balance between the two fluid bed reactors a bed material stream is drained off from the combustor's dense bed and returned to the pyrolyzer. The fines from the combustor's ideal gas-solid separator are returned to the combustor. The difference in pyrolysis heat demand and exothermic energy supply of the char combustion reaction is balanced by transferring excess heat to a cooling medium in the fluidized bed combustor ($\dot{Q}_{fbc}$). Depending on the operation parameters the heat flow $\dot{Q}_{fbc}$ might also be reversed. This reversal in heat flow can e.g. be envisaged as an additional heat transfer from the pyrolysis gas burner into the recirculation system. In the ideal gas-oil separator the pyrolysis reaction is quenched and the pyrolysis vapors condensed. The pyrolysis gas is combusted in the ideal gas burner. The gas supply for all reactors is preheated. Simulated is Kraft lignin pyrolysis with the input parameters listed in Table 5.11.

5.4.3 Energetic process performance ratios

For performance evaluation of pyrolysis process, the energy recovery rate of the liquid product η_{OR} is employed. It is defined as the ratio of fuel energy of pyrolysis oil and of lignin feed. The oil mass flow can be determined by $\dot{m}_{oil,dry} = Y_{oil,dry} \cdot \dot{m}_L$. The higher

Table 5.11: Input parameters for simulation of the integrated pyrolysis–combustion process

parameter	unit	pyrolyzer (riser)	combustor (BFB)
pressure	bar	1.2	1.2
temperature	°C	400 to 700	850
superficial gas velocity	m/s	2.5 to 5	0.2 to 1
biomass feed	kg/h	5, 10, 15, 20	–
biomass composition[†]	wt.-%	100 (lignin)	–
excess air ratio	–	-	1.3
solids inventory of bed	kg	0.75	5
sauter diameter sand	µm		200	
reactor height	mm	2000	900
reactor diameter	mm	80	200, 400[‡]

[†]correlates to Kraft lignin composition, [‡]at $\dot{m}_L = 15$ and $20\,\mathrm{kg/h}$

heating values $H_{0,\mathrm{oil}}$ and $H_{0,\mathrm{L}}$ are taken from Figure A.1 and Table 4.4 (Kraft lignin), respectively.

$$\eta_{\mathrm{OR}} = \frac{\dot{m}_{\mathrm{oil,dry}} \cdot H_{0,\mathrm{oil}}}{\dot{m}_{\mathrm{L}} \cdot H_{0,\mathrm{L}}} = Y_{\mathrm{oil,dry}} \cdot \frac{H_{0,\mathrm{oil}}}{H_{0,\mathrm{L}}} \tag{5.57}$$

Based on a surplus-deficit-ratio defined by Boukis et al. [88] an energetic performance ratio for the coupled pyrolyzer-combustor-process is derived. At values above 1 it can be expected that the energy supply by combustion of both char and pyrolysis gas minus the heat required for preheating of the gases is greater than the demand for the endothermic pyrolysis reaction and heat losses.

$$\eta_{\mathrm{SD}} = \frac{\dot{Q}_{\mathrm{R,comb}} + \dot{Q}_{\mathrm{R,gas\ comb}} - \dot{Q}_{\mathrm{gas}}}{\dot{Q}_{\mathrm{R,PR}} + \dot{Q}_{\mathrm{loss}}} \tag{5.58}$$

$\dot{Q}_{\mathrm{R,comb}}$, $\dot{Q}_{\mathrm{R,gas\ comb}}$ and $\dot{Q}_{\mathrm{R,PR}}$ are the heat released by combustion of char and permanent gases as well as the heat required for endothermic pyrolysis, respectively. Each combustion heat release can be calculated by the product of the higher heating value of gas and char (cf. Figure A.1) times the corresponding mass flow. The pyrolysis energy requirement for Kraft lignin is given by the product of $\Delta H_{\mathrm{R,pyr}}$ (Figure A.2a) and the feeding rate of lignin. Furthermore, the overall heat flows for preheating of the gases $\dot{Q}_{\mathrm{gas}}$ is

$$\dot{Q}_{\mathrm{gas}} = \dot{Q}_{\mathrm{GH,PR}} + \dot{Q}_{\mathrm{AH,comb}} + \dot{Q}_{\mathrm{AH,gas\ comb}} + \dot{Q}_{\mathrm{GH,gas\ comb}} \tag{5.59}$$

and the heat loss, which is approximated by 0.1 times the circulating enthalpy flow between the reactors (analogously to the definition of Boukis et al. [88]), is defined as:

$$\dot{Q}_{\mathrm{loss}} = 0.1 \cdot (\dot{H}_{\mathrm{PR}\rightarrow\mathrm{comb}} + \dot{H}_{\mathrm{comb}\rightarrow\mathrm{PR}}) \tag{5.60}$$

6 Pyrolysis results

The outcome of the CFB pyrolysis experiments for Kraft and hydrolysis lignin are discussed in this chapter. Starting with the pyrolysis mechanism in the CFB system, the differences between the mechanisms for both lignins are illustrated. The discussion is continued with the product distribution of CFB lignin pyrolysis. Therefore, recovery rate and possible errors are evaluated, before the yield and composition of gas, oil, and char are examined. It is focused on the influence of temperature and mineral matter in char.

6.1 Pyrolysis mechanism and particle size distribution

The morphology and particle size distribution of the solid residue of solid fuel pyrolysis gives important information about the devolatilization behavior [301] and thus the pyrolysis mechanism. The cumulative particle size distributions of both bed and secondary cyclone material are shown together with the distribution of quartz sand in Figure 6.1a and 6.1b for Kraft and hydrolysis lignin pyrolysis, respectively. The secondary cyclone material distributions are both bimodal and the bed materials are of larger size than the quartz sand. The only obvious deviation between the two different lignin types is the much larger coarse fraction in the bed material from hydrolysis lignin pyrolysis. Also, a larger maximum size of this bed material of up to 1500 µm can be observed.

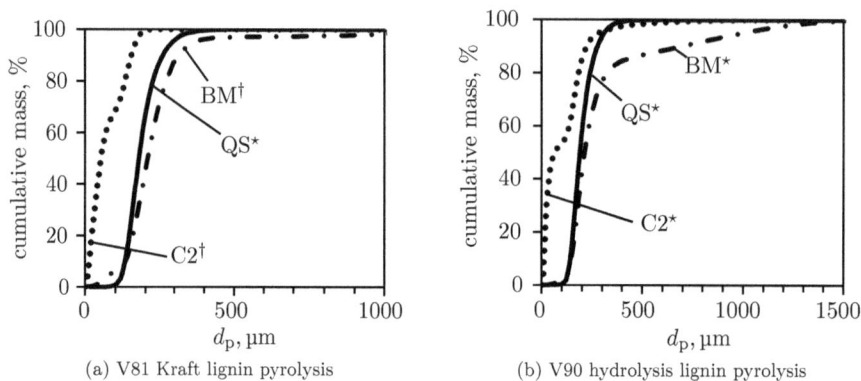

(a) V81 Kraft lignin pyrolysis   (b) V90 hydrolysis lignin pyrolysis

Figure 6.1: Cumulative mass distributions of bed material (BM) and secondary cyclone material (C2) compared to quartz sand (QS)

measured with: *Camsizer XT, †Beckman Coulter

The SEM images in Figure 6.2 show the materials of the Kraft lignin pyrolysis process. The bed material consists of mainly coarse particles with similar shape to quartz sand (cf. Figure 4.20) but also some small plate-like particles of much smaller size (Fig. 6.2a). In contrary, the secondary cyclone material (Figure 6.2b) mainly consists of the plate-like particles and few coarse bed material particles. These two particle types induce the

(a) Bed material (V78BM)

(b) Secondary cyclone material (V81C2)

(c) V81BM particle

(d) V81C2 char coating fragments

Figure 6.2: SEM images of bed and secondary cyclone material from Kraft lignin pyrolysis

bimodal mass distribution in Figure 6.1a (also the few fines in the bed material). At higher resolution, it can be observed that the bed material particles are coated with char during pyrolysis. A fracture of the coating is shown in Figure 6.2c. The char surface shows no macro pores. The specific surface area measurement for the char fragments (Figure 6.3) reveals that meso and micro pore area have their maximum of about 130 and $500\,m^2/g$ at a pyrolysis temperature of $650\,°C$, respectively. The meso pore area is smaller than $2\,m^2/g$ at $550\,°C$. The micro pore area (average pore width: $0.8\,nm$) is larger than $135\,m^2/g$ for all temperatures. It should be mentioned at this point, that the particle shape has an influence on the surface area as well. The Kraft lignin char fragment particles have a more plate-like shape. At given volume and the present particle dimensions, the surface ratio of plates and a sphere is roughly 5 to 15. Using this approximation, the specific surface area for spherical particles would be in the range of 2 to $26\,m^2/g$ and 9 to $100\,m^2/g$ for meso and micro pores, respectively. This approximation is in good agreement with the BET area of $< 5\,m^2/g$ from Kraft lignin pyrolysis at 250 to $550\,°C$ [123].

The char coating fragments are mainly in contrast to the coated quartz sand particles to a large extend not caught in the primary cyclone and therefore found in the secondary cyclone material (Figure 6.2d). Additionally, small spherical particles are found in the secondary cyclone material (arrow in Figure 6.2d). SEM-EDX measurements reveal that these spherical particles consist of mainly carbon and oxygen. Thus, it can be concluded that they are char particles. Dimensioning of these spherical char particles in SEM images shows that the particle diameter is mainly in the range of 1.6 to $30.3\,µm$ with a

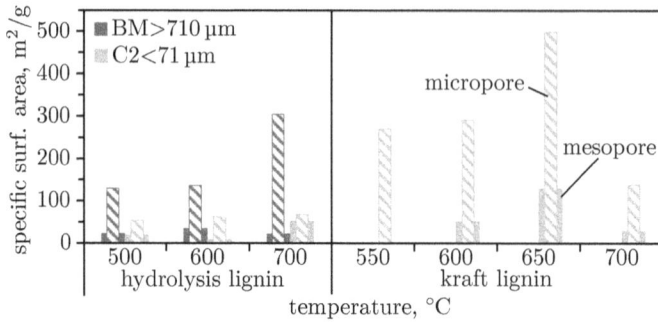

Figure 6.3: Micro- (dashed) and mesopore (solid) specific surface area of bed material (BM) and secondary cyclone material (C2) fractions (sieved) characterization according to [251, 302]: micro pores $d_{pore} < 2\,$nm, meso pores $2\,$nm $< d_{pore} < 50\,$nm, macro pores $> 50\,$nm

mean value of 13.8 µm. Thus, these spherical particles are either in the size of or slightly larger than primary lignin particles in Figure 4.15a.

For hydrolysis lignin pyrolysis a different pyrolysis result arises. There are two types of particles in the bed material. First particles originating from the initial bed material (quartz sand) and second particles with high porosity (Fig. 6.4a). Figure 6.4b shows SEM images of the secondary cyclone material at two resolutions. At lower magnification (Fig. 6.4b, left) one quartz sand and small pyrolysis product particles can be seen. This morphology is the reason for the bimodal distribution in Figure 6.1b. At higher magnification (Fig. 6.4b, right) it can be seen that the fines in the C2 material are much more irregular but overall closer to spherical shape than the plate-like char fragments resulting from Kraft lignin pyrolysis. Also, twig-like particles can be seen, which result from pyrolysis of the twig-like particles in the hydrolysis lignin (cf. Figure 4.17b). Furthermore, some particles appear to have broken in the process (not depicted). The surface of the particles originating from the initial bed material quartz sand (cf. Fig. 6.4a) is shown in Figure 6.4c, right. Compared to the initial quartz sand surface (Fig. 6.4c, left) it is recognizable that the particle is not coated with a thick and homogeneous char layer, but fine char particles adhere to the quartz sands surface. The surface of the char particle with high porosity shows macro pores in the size range of about 0.1 to 50 µm; the pores are larger at higher pyrolysis temperature (depicted in Figures 6.4d, left for 500 °C and right for 700 °C). The meso and micro pore specific surface areas (cf. Fig. 6.3) of the large porous bed material particles ($d_p > 710\,$µm) are 23 to 35 m^2/g and 130 to 305 m^2/g, with average pore sizes of 6 to 12 nm and 1.4 to 2.2 nm, respectively. The micro pore volume increases from 0.048 to 0.114 cm^3/g. Thus, the specific surface area, pore width, and pore volume of the micro pores increase with temperature. This trend holds also for the C2 material particles. The C2 material particles of $d_p < 71\,$µm have a meso and micro pore specific surface area in the range of 9 to 50 m^2/g and 53 to 67 m^2/g, respectively. The corresponding average pore widths are 8 to 21 nm and 1.2 to 1.4 nm. The micro pore volume increases from 0.02 to 0.025 cm^3/g. An explanation for the above-mentioned increase of specific surface area with increasing pore width could be that with the increase of and more rapid devolatilization, at higher temperatures, the overall porosity and pore volume increases. The specific meso pore surface area is in the magnitude of switch grass and corn stover char, obtained by fast pyrolysis in a bubbling fluidized bed at 500 °C, with 7 to 22 m^2/g [207] and hardwood chars with 2 to 100 m^2/g

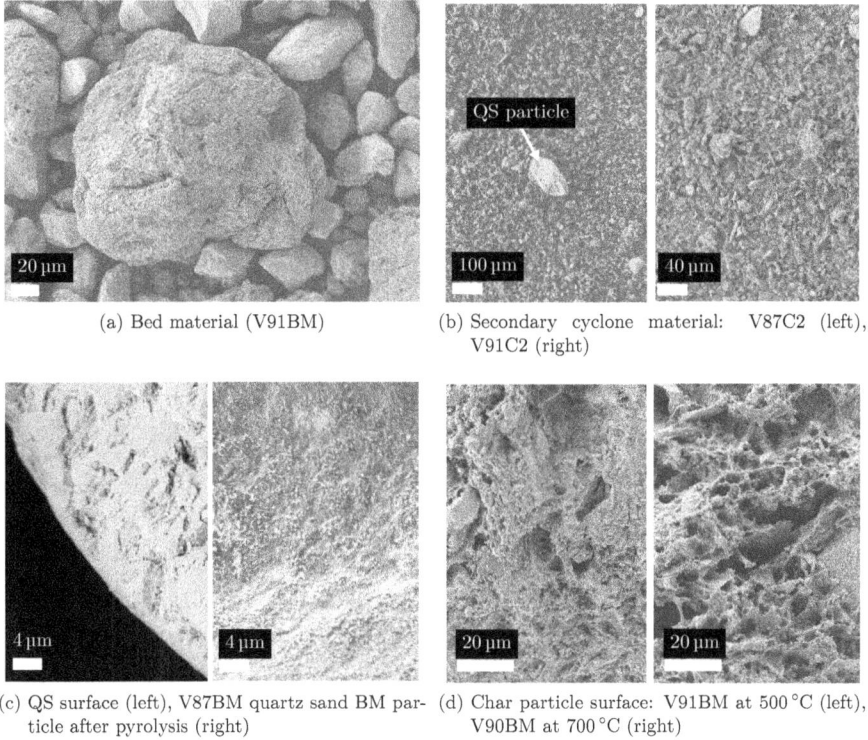

(a) Bed material (V91BM)

(b) Secondary cyclone material: V87C2 (left), V91C2 (right)

(c) QS surface (left), V87BM quartz sand BM particle after pyrolysis (right)

(d) Char particle surface: V91BM at 500 °C (left), V90BM at 700 °C (right)

Figure 6.4: SEM images of bed and secondary cyclone material from hydrolysis lignin pyrolysis

[303]; but lower than for steam activated softwood char with 450 to 610 m^2/g (activation at 600 to 985 °C) [304]. The micro pore surface area is also in the magnitude of literature data: for example, wood char from fluidzed bed pyrolysis of eucalyptus was determined to have a surface area of 540 m^2/g (pyrolysis at 800 and 900 °C) [305] and hardwood chars 180 to 510 m^2/g [303]. Commercially activated carbon used for liquid and gas phase adsorption processes has meso and micropore surface areas of 800 to 1400 m^2/g and 800 to 1200 m^2/g [302]. The here obtained values are slightly smaller but bigger than or in the magnitude of molecular sieves [302]. To enlarge the specific surface area of the char, of which especially the hydrolysis lignin char with $d_p > 710$ μm might be interesting for adsorption applications, steam or CO_2 activation might be an option [306].

As shown in Section "Thermal characterization" 4.4.1, at high heating rates lignins with high purity melt (Kraft and organosolv lignin), whereas lignins with considerable cellulose and hemicellulose content (hydrolysis and soda lignin) keep their particulate structure. Thus, it can be concluded that the major difference during pyrolysis of the two different lignins used for CFB pyrolysis is the melting of Kraft lignin, whereas melting does not occur for hydrolysis lignin with higher impurity. Furthermore, it was observed that more fine char particles have been entrained from the CFB system, i.e. haven't been retained by the primary cyclone, for pyrolysis of hydrolysis lignin. This observation is in good agreement with the other findings as the fine particles do not melt and stick to the bed material surface and have thus a higher probability not to be separated by a cyclone.

With the gained information, pyrolysis schemes for Kraft and hydrolysis lignin, depicted in Figures 6.5 and 6.7, respectively, are derived.

The Kraft lignin particles enter the riser reactor at low temperature through the cooled lance. In the bed the particles are quickly heated by convection, radiation and if a contact incident occurs conduction. Thus, the lignin is melting in a temperature range of 215 to 285 °C (cf. Section "Thermal characterization" 4.4.1). The molten lignin either spreads on a quartz sand particle or reacts completely in a shape of a liquid droplet. The latter is happening when no contact incident occurs before the lignin surface is reacted far enough to build a non-sticky char surface. Therefore, resulting in the generation of spherical particles (as shown in Fig. 6.2d), which is also justifying this pyrolysis scheme derivation. However, the number of spherical particles is very low compared to the char layer fragment and coated bed material particle numbers. The main fraction of the lignin particles impact on bed material particles and melt on their surface, reacting either successively or in parallel building a char layer. This process is faster than the sole reaction of a lignin particle. Due to the spreading of molten lignin on a hot quartz sand particle a large contact area forms, resulting in fast heat transfer by conduction. From the Flash-DSC measurements and from the surface morphology of the Kraft lignin char, which are virtually macro pore free (cf. Fig. 6.2c and Fig. 6.6 – material and apparent densities are almost equal for Kraft lignin char), it can be concluded that the main devolatilization takes place while the intermediate layer is molten. Hence, if a volatiles bubble (cf. droplet picture in Figures 4.18(a) and 4.19(d) for Kraft and organosolv lignin, respectively) is released from the liquid layer, the surface tension evens out the surface. After solidification, the char layer is not stable when subjected to mechanic stress due to particle impact on the reactor wall, cyclone wall or collision with other particles. Thus, giving birth to char layer fragments and attrition fines. [93]

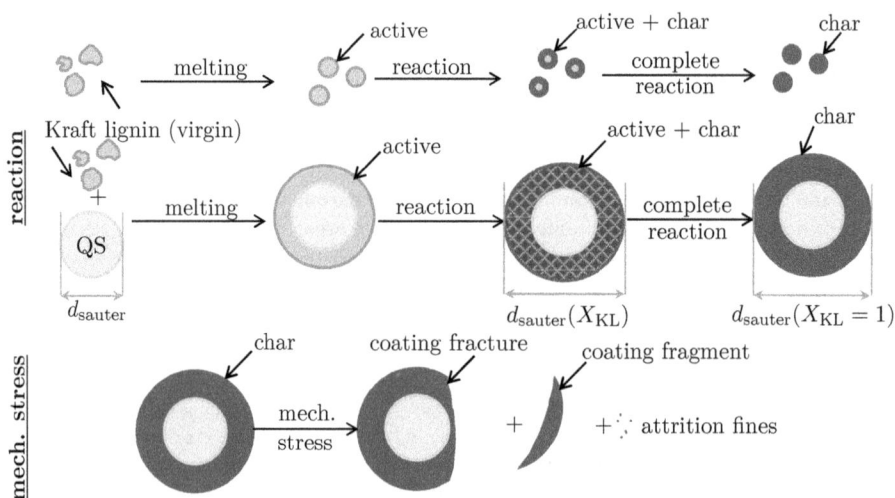

Figure 6.5: Pyrolysis mechanism of Kraft lignin

The picture is different for pyrolysis of hydrolysis lignin. The lignin particles react without melting. Therefore, as gases evolve through rapid reaction, pores are formed while the overall particle shape remains. A measure for the pore evolution is the particle porosity, which is defined as:

$$\varepsilon_p = 1 - \frac{\rho_{a,char}}{\rho_{m,char}} \quad . \tag{6.1}$$

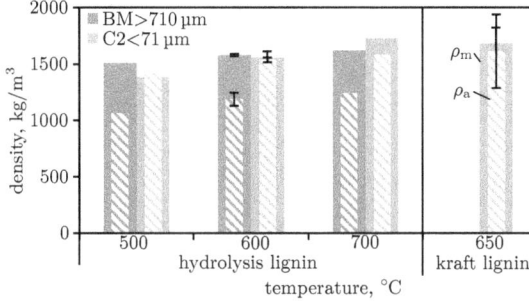

Figure 6.6: Densities of bed material (BM) and secondary cyclone material (C2) fractions (sieved), material density ρ_m (solid), apparent density ρ_a (dashed) calculated by equation 4.2

Through the mass loss by devolatilization (and in some extent breakage and attrition), the particles shrink. For the coarse char particles with $d_p > 710\,\mu m$ (virtually quartz sand-free), the shrinkage (neglecting breakage and attrition) can be approximated by the measured apparent densities of lignin $\rho_{a,lignin}$ and char $\rho_{a,char}$ together with the char yield Y_{char}. It holds:

$$S = 1 - \frac{V_{char}}{V_{lignin}} = 1 - \frac{\rho_{a,lignin}}{\rho_{a,char}} \cdot Y_{char} \quad . \tag{6.2}$$

The char densities in Equations 6.1 and 6.2 have been obtained for the sieved samples consisting mainly of char and are depicted in Fig 6.6. The results for hydrolysis lignin are listed in Table 6.1. Both shrinkage and porosity increase with rising temperature due to a higher mass loss. The higher porosity at rising temperature indicates a higher pore volume, which can also be seen in Figure 6.4d. The more intense pore formation at higher temperature can be explained by a faster and intenser pyrolysis reactions with formation of more volatiles and rapid release of these volatiles from the solid matrix. When compared to the char surface area of Kraft lignin (converted by the factor 5 to 15 from char layer fragments to spherical shape) the specific surface area of the hydrolysis lignin char (with $d_p > 710\,\mu m$) is larger. As the hydrolysis lignin does not melt, the quartz sand is not coated homogeneously by char. But the quartz sand surface is covered, presumably due to adhesive forces, by small particles in the scale of 10 to 1000 nm (cf. SEM pictures). This particle size is in the relevant order of magnitude for van-der-Waals forces. Also, solid bridges might have formed due to partial stickiness. Due to mechanic stress in the reactor or cyclone some particles break and attrition generates fines. This comminution is presumably promoted by the char porosity.

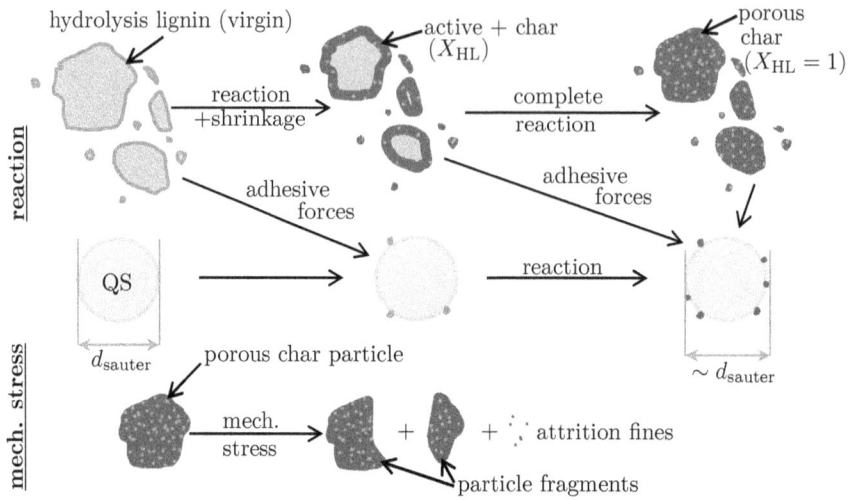

Figure 6.7: Pyrolysis mechanism of hydrolysis lignin

The overall conclusion that pure lignin melts, leading to a granulation of bed material, whereas higher impurity with cellulose and hemicellulose prohibits melting and hence hemicellulose lignin reacts to form porous char particles is also supported by evidence found in literature. SEM imaging of the silica sand bed material after fluidized bed pyrolysis at 550 °C of corncob (12 to 18 wt.-% lignin content [307, 308]) in N_2-atmosphere revealed that only some small char particles stick on the bed material surface [156]. Also, during pyrolysis of beech wood (about 23 wt.-% lignin [309]) in a fluidized bed, it was found that the biomass particles do not melt but shrink due to release of volatiles [310]. For Kraft lignin on the other hand Sharma et al. [123] showed by SEM images that it softened, molten and fused into a matrix of particles (solid bridge agglomeration) below 250 °C. The pyrolysis of HDPE (melting point between 120 and 180 °C) in a fluidized bed showed also that the sand used as bed material is coated with char. Thus, as shown in this work, the fast fluidized bed system can be applied to lignin pyrolysis. Although agglomeration was observed at ≤ 500 °C, at higher temperatures granulation prevails. The reported problems in bubbling fluidized beds (cf. 2.2.2) can be prevented and mechanical stirring of the fluidized bed [85] is not necessary. It can be concluded that at higher pyrolysis temperature the reaction is fast enough to form a low sticky lignin-char-intermediate layer concentration, which together with the more vigorous particle movement and lower particle concentration prevent agglomeration.

Table 6.1: Hydrolysis lignin: particle shrinkage S due to devolatilization and final char porosity ε_p for bed material particles with $d_p > 710\,\mu m$

ϑ_{pyr}	$S^\dagger$	$\varepsilon_p^\star$
°C	vol.-%	–
500	76.3	0.18
600	76.9	0.25
700	78.9	0.35

calculated by †Eq. 6.2 and *Eq. 6.1

6.1.1 Granulation

Fluidized bed granulation has been investigated mainly in the stationary fluidized beds [296]. Only a few studies exist on the granulation behavior in circulating fluidized bed systems [311–318]. It was found that in the CFB system the agglomeration affinity is reduced compared to bubbling fluidized beds [313] and can be explained by the more vigorous particle movement and the lower impact probability in the fast fluidized system. Furthermore, it was found that granulation can be carried out at lower temperature [315] in fast fluidization compared to bubbling fluidized bed systems. Additionally, Stiller et al. [316] investigated the granulation behavior of sucrose in a circulating fluidized bed. They found that the increase of injected droplet size (for lignin the feed particle size) and injection mass flow increase the particle growth rate, whereas the growth rate is decreased by increasing fluidization velocity (breakage). Furthermore, they found that with the decrease of bed material holdup the liquid-to-solid ratio in the bed is increased, which leads to agglomeration.

The granulation rate of the quartz sand particles with char depends on the amount of lignin fed into the reactor during an experiment and the corresponding char yield. For experiment V81 the char layer thickness was approximated by dimensioning of the breakage cross-section of more than 70 char layer fractures and char layer fragments in SEM images giving a range of approximately 0.46 to 18.1 μm with a mean value of 6.7 μm. This result is in good agreement with the bed material particle size distributions shown in Figure 6.8. For the char layer thickness, a rough estimate is the half difference between the bed material and the quartz sand Sauter diameter giving 7 and 18 μm for BMV57 and BMV81, respectively.

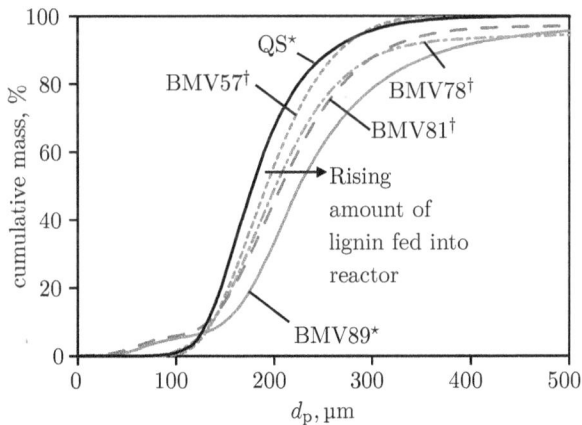

Figure 6.8: Cumulative mass distributions of bed material (BM) depending on lignin mass fed into the reactor

measured with: *Camsizer XT, †Beckman Coulter

A better estimation is achieved, when calculating the char layer thickness s from the Sauter diameter of the char layered bed material $d_{sauter, BM}$ and the volume fraction of the char φ_{char} in the bed material sample. It holds:

$$s = 0.5 \cdot d_{sauter, BM} \cdot \left(1 - \sqrt[3]{1 - \varphi_{char}}\right) \qquad . \tag{6.3}$$

The volume fraction of char φ_{char} can be calculated from the measured apparent densities of char $\rho_{\text{a, char}}$ and quartz sand $\rho_{\text{a, QS}}$ and the char mass fraction w_{char}.

$$\varphi_{\text{char}} = \frac{V_{\text{char}}}{V_{\text{p}}} = \frac{w_{\text{char}}/\rho_{\text{a, char}}}{w_{\text{char}}/\rho_{\text{a, char}} + (1 - w_{\text{char}})/\rho_{\text{a, QS}}} \tag{6.4}$$

The char layer thickness increases with feeding time and thus the mass of char produced and deposited on the bed material surface (Figure 6.9b). The particle growth rate [296] may be defined by the following equation. For the Kraft lignin pyrolysis particularly, the right-hand side holds:

$$\frac{\mathrm{d}d}{\mathrm{d}t} = \frac{2 \cdot \dot{m}_{\text{char}}}{\rho \cdot A_{\text{total}}} = \frac{2 \cdot \dot{m}_{\text{L}} \cdot Y_{\text{char}}(t)}{\rho_{\text{a, char}} \cdot \pi \cdot d^2_{\text{sauter}} \cdot N_{\text{p}}} \quad . \tag{6.5}$$

Inserting the linear dependency from Figure 6.9a for $Y_{\text{char}}(t)$ provides the solid lines in Figure 6.9b, respectively. Model parameters are average parameters for the evaluated experiments: $\dot{m}_{\text{L}} = 2.2\,\text{kg/h}$, $\rho_{\text{a, char}}$ from Figure 6.6, $d(t_0) = d_{\text{sauter, QS}} = 160\,\mu\text{m}$ and the number of particles in the bed $N_{\text{p}} = m_{\text{QS, bed}}/(\pi/6 \cdot d^3_{\text{sauter, QS}} \cdot \rho_{\text{a, QS}}) = 3.6 \cdot 10^8$. It can be concluded that the char growth rate does increase due to the rise in char yield with rising amount of char in the reactor.

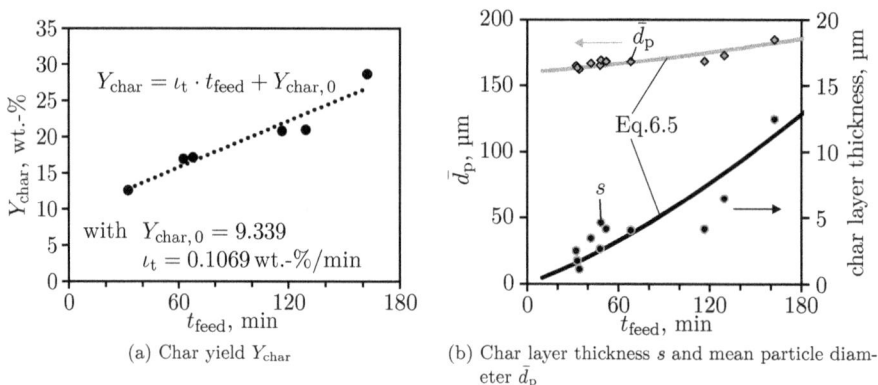

(a) Char yield Y_{char}

(b) Char layer thickness s and mean particle diameter $\bar{d}_{\text{p}}$

Figure 6.9: Dependency of char yield and char layer build-up on feeding time t_{feed}

For further investigation, the bed materials from V57, V81, and V89 were sieved in five fractions each. The size classes are listed in Table 6.2. The particle size distribution of each fraction were then measured. Exemplary, the resulting cumulative particle size distributions for BMV89 are shown in Figure 6.10. In this particular case, some char layer fragments (bimodal distribution of cl. 1) and partial agglomeration (cl. 5) can be observed. In BMV57 no char layer fragments and agglomerates are found, while only a few char layer fragments are found in BMV81. From the particle size distributions the Sauter diameters were calculated and additionally the char content measured. With Eq. 6.3 the char layer thickness of each class and the overall bed material sample are determined. Figure 6.11 shows the result. It can be seen that with rising sauter diameter (cl. 1 to cl. 5) and rising amount of lignin fed into the reactor the char layer thickness increases. This increase holds for both sieved sample fractions and the whole sample (mean value, empty circle in Figure 6.11). Furthermore, this increase is linearly depending on particle size. Only the smallest and largest size fractions of BMV89 do deviate from this linear trend as explained above by partial content of char layer fragments and agglomeration.

Table 6.2: Particle size classes for determination of particle size depending char layer thickness

class no.	BMV57	BMV81	BMV89
	particle size class, μm		
cl. 1	<140	<160	<180
cl. 2	140-180	160-200	180-200
cl. 3	180-200	200-224	200-224
cl. 4	200-224	224-250	224-280
cl. 5	>224	>250	>280

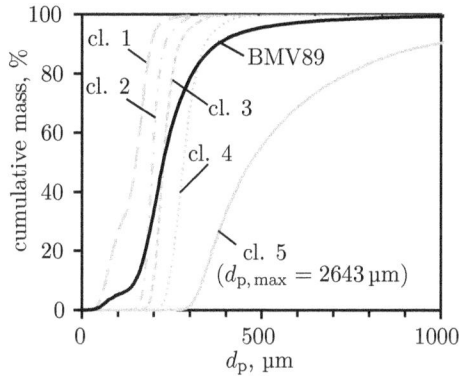

Figure 6.10: Cumulative particle size distributions of sieved fractions for BMV89

A reason for this growth behavior could be the higher heat capacity of the larger particles leading to more granulation. Furthermore, the inertia of bigger particles causes a longer residence time of the larger particles in the spraying zone and thus a growth rate, which is increasing with rising particle diameter. This is in good agreement with findings of Zank [319], who found that larger particles might have higher recirculation frequencies into the spraying zone, which also increase the growth rate of particles with larger size. Furthermore, it was found that segregation occurs in circulating fluidized beds. Hirschberg and Werther [320] found that this segregation effect mainly depends on the terminal particle velocity, which itself reflects the influence of particle size and density. Therefore, larger particles (or particles with higher density) with a higher terminal particle velocity have a higher probability to be in the lower part of the CFB riser and thus also a higher probability to be coated with char. A further effect could be the embedding of char attrition fines, which are retained in the CFB system by the cyclone, back into the char layer (cf. also [318]). This embedding can happen when such a fine particle impacts on the sticky lignin-char-intermediate layer surface.

Thus, it can be concluded that the char layer growth rate of the particle size fractions is not governed by a surface proportionality as larger size fractions grow faster than smaller size fractions. However, including the yield dependency, the average growth rate can sufficiently be described by Eq. 6.5.

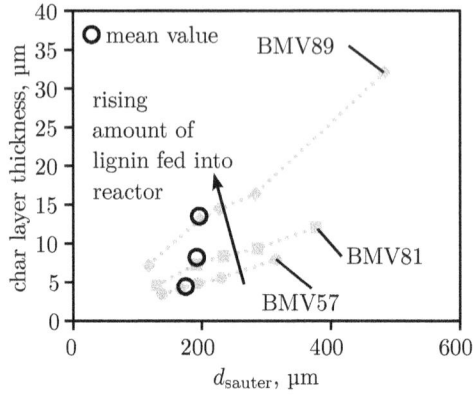

Figure 6.11: Particle size depending char layer thickness
calculated by Eq. 6.3

6.2 Influence of pyrolysis parameters

6.2.1 Pyrolysis temperature

The pyrolysis temperature is one of the major parameters influencing the product distribution [56]. In the temperature range of 550 to 700 °C and 500 to 700 °C, the resulting yields of pseudo-components char, oil, and gas as well as the composition of pyrolysis oil and gas are calculated for Kraft and hydrolysis lignin pyrolysis. Furthermore, the recovery rate and the nature of occurring losses are discussed.

6.2.1.1 Pseudo-product distribution

The yields of the pseudo-components char, oil, and gas in dependence of reactor temperature are shown in Figure 6.12. The depicted oil yield, which is determined according to Eq. 4.18, does not include water. In the considered temperature range this oil yield is higher for Kraft lignin than for hydrolysis lignin. The optimum is 49 wt.-% at 600 to 650 °C. For hydrolysis lignin the maximum is 36.5 wt.-% at 500 °C. But, as the maximum hydrolysis oil yield was found at the minimal pyrolysis temperature, it should be mentioned that the global oil yield maximum could be at a slightly lower temperature. With rising temperature, the oil yield of hydrolysis lignin decreases to 10.6 wt.-% at 700 °C. The difference can be explained by the composition of the two lignins. Kraft lignin has a purity of 95.4 wt.-%, which is much higher than the purity of hydrolysis lignin with a carbohydrate content of 43.8 wt.-%. Carbohydrates decompose at lower temperature range than lignin. Therefore, the temperature optimum of hydrolysis lignin is shifted relative to Kraft lignin to lower temperature. Furthermore, the water content of the hydrolysis lignin oil is with 25 wt.-% higher than that of Kraft lignin (10 wt.-%). Thus, the difference between the total liquid yield (water and oil) is not as high as for the total oil yield depicted. The reason for the higher water content of the liquid product

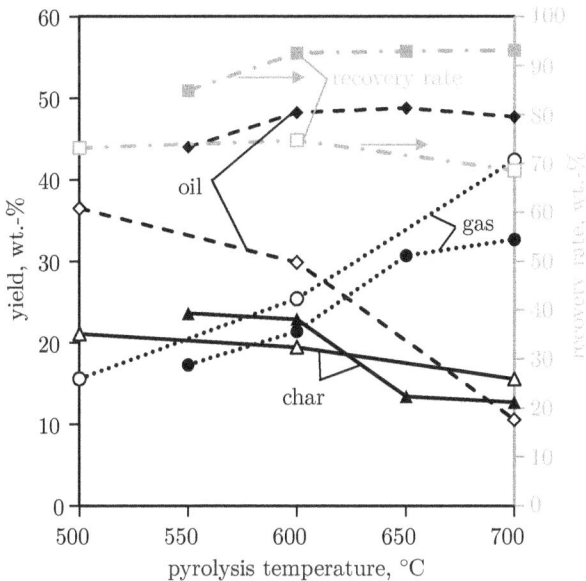

Figure 6.12: Yields of pyrolysis products and recovery rate for Kraft lignin (filled symbols) and hydrolysis lignin (empty symbols)

of hydrolysis lignin pyrolysis is that the carbohydrate decomposition reactions result in a higher water yield. Thus, the obtained oil yields are mostly higher than in other pyrolysis equipment such as fixed bed or centrifuge reactor and at the top of the range reported for fluidized bed and entrained flow lignin pyrolysis (cf. Table 6.3).

Table 6.3: Liquid yields of lignin pyrolysis in technical reactors

lignin type	reactor type[†]	temp. °C	oil yield (wt.-%, daf) with reaction water	without	ref.
straw lignin	CPR	550	39	31	[124, 163]
Alcell	FiBR	570	23	-	[117]
Kraft	FiBR	610	17.5	-	[115, 117]
ETEK	FBR	500	57.7	45.3	[48]
ETEK	FBR	450-510	72	-	[48]
ETEK	FBR	484-519	40.1	-	[48]
ETEK	FBR	500	47	-	[48]
ETEK	FBR	480	40	-	[48]
ETEK	EFR	700	11.7	-	[48]
ALM	FBR	530	31.2	19.8	[48]
ALM	FBR	450-510	31	-	[48]
ALM	FBR	475-525	49.7	-	[48]
ALM	FBR	480	22	-	[48]
ALM	EFR	700	36.6	-	[48]
Granit	FBR	500	47.6	-	[29]
Alcell	FBR	500	38.9	-	[29]
A, Organosolv	FBR	500	54.7	-	[29]
B, Organosolv	FBR	500	51.9	-	[29]
Indulin AT	FBR	550	23	-	[125]
Lignoboost	FBR	550	22	-	[125]
Acetocell	FBR	550	16	-	[125]
Kraft	MFBR	550	45	-	[85]
Kraft	CFBR	600-650	60	49	this study
Hydrolysis	CFBR	500	60	36.5[‡]	this study

[†]CPR: centrifugal pyrolysis reactor, FBR: fluidized bed reactor, MFBR: mechanically fluidized bed reactor, FiBR: fixed bed reactor, CFBR: circulating fluidized bed reactor, EFR: entrained flow reactor; [‡]possibly higher at lower temperature

Between 500 and 700 °C the char yield of hydrolysis lignin decreases from 21.1 to 15.6 wt.-%, whereas the char yield of Kraft lignin decreases from 23.5 to 12.7 wt.-% in the range of 550 to 700 °C. The char yield is influenced by both the carbohydrate composition and the ash content of the pyrolyzed biomass. Although the final char yield of cellulose and hemicellulose is considerably lower than that of Kraft lignin (cf. Fig 2.7), the char yield for both lignins has the same magnitude. This same magnitude can be explained by the substantially higher ash content of hydrolysis lignin compared to Kraft lignin, i.e. 11.3 wt.-% compared to 1.1 wt.-%, respectively.

Both lignin gas yields increase with temperature due to more intense cracking at rising temperature. In the considered temperature range the gas yield increases from 15.6 to 42.4 wt.-% and from 17.2 to 32.7 wt.-% for hydrolysis and Kraft lignin, respectively. The total gas yield of hydrolysis lignin pyrolysis is 4 to 10 wt.-% higher than that of Kraft lignin. This difference can also be attributed to the higher carbohydrate content of the hydrolysis lignin and is discussed in detail below.

6.2.1.2 Recovery rate and error discussion

Figure 6.12 additionally shows the recovery rate, which is defined as

$$X_{RR} = Y_{char} + Y_{oil} + Y_{gas} \qquad . \tag{6.6}$$

It can be seen that the recovery rate for the experiments with Kraft lignin is with 84.7 to 93.1 wt.-% up to about 20 wt.-% higher than that of the hydrolysis lignin experiments. Possible losses not included in the mass balance (cf. Section 4.3) might be solid, liquid or gaseous products. Solid char that is entrained from the secondary cyclone is not considered by the material balance. The entrained solids enter the scrubber where they are retained in the liquid phase, as downstream in the electrostatic precipitator and heated filter of QIR2 (cf. Figure 4.6) no solids were found. For determination of solids found in the scrubber, the scrubber sump was analyzed. After pyrolysis usually two immiscible phases developed. The aqueous phase was filtered, whereas the tarry phase was Soxhlet extracted with acetone. For the Soxhlet extracted phase both the residue of the extraction and the oil-acetone solution filter residue was analyzed. During few Kraft lignin pyrolysis experiments some char particles were found in the scrubber sump. But during most experiments, the tarry phase totally dissolved in acetone and in the aqueous phase no particles were found. From the scrubber sump of hydrolysis lignin pyrolysis, solids with a yield of in total 1 to 3 wt.-% could be found. The Soxhlet extraction residue of the tarry scrubber phase has the composition of char. This finding indicates that some char was indeed entrained and retained in the tarry scrubber phase for hydrolysis lignin. The filter residue of the aqueous phase had typical brown (Kraft) lignin color and an intermediate composition between oil and char. SEM images revealed that the filter residue consists mainly of spherical particles with particle size of 0.5 to 10 μm. Hence, it can be concluded that some liquid product called pyrolytic lignin precipitates in the scrubber sump to form these solidified droplets. Further information on pyrolytic lignin can be found elsewhere [46, 188, 321].

For determination of gas yield and composition both online measurements and gas sample bags were taken. In most instances, there is a good agreement between both methods. But H_2 is only measured for some experiments (detailed discussion below) leading to a recovery rate reduction of less than 1 wt.-%. The overall balancing and gas analytic procedure was validated by two experiments at pyrolysis conditions (650 °C) without lignin feeding, but with analogous sampling. By adding a known amount of tracer gas (CO_2) to the fluidizing gas and determination of the CO_2-recovery rate it was shown that the gas balance can be determined within an error of 6 % [322].

Also, the overall oil balancing scheme (cf. Section 4.3.2) was validated by the two experiments without lignin feeding. Next to the above-described CO_2-addition, also a known amount of water was injected into the circulating fluidized bed. The downstream sampling with subsequent KFT analysis revealed that the oil yield can be determined with an error of about 20 % [322]. The error is larger than for gas because of the relatively small sampling rate of about 0.5 %. During liquid sample characterization of lignin pyrolysis, both the distillate from rotary evaporation (path 1 & 2b) and distillation (path

2a) in Figure 4.10 showed no detectable amount of pyrolysis products (determined by head space GC). Oil sample preparation (cf. steps in Figure 4.10) and subsequent GC analysis were tested and enhanced by adding exactly known amounts of both typical phenolic decomposition products and the internal standard (fluoranthene) to an isopropanol sample and successive three-point calibration. With the determined recovery rates for the single decomposition products the preciseness of chromatographic quantification could be improved [322]. But due to the chosen analysis procedure, not all low boiling alcohols (most importantly CH_3OH) – due to the lower boiling point of CH_3OH compared to isopropanol – might be accounted for; contributing to an incomplete mass balance with an error in the magnitude of 1 to 2 %. For example, Nunn et al. [126] found the methanol yield of MWL pyrolysis to be in the magnitude of 1.5 wt.-%. The water content of the pyrolysis oils can only be determined for experiments where solely N_2 is used as the fluidizing medium. The water content was determined for the hydrolysis lignin pyrolysis experiments V88 (600 °C) and V91 (500 °C) as well as for Kraft lignin pyrolysis experiment V89 (650 °C). The resulting yields of water are 28.1 wt.-%, 25.2 wt.-%, and 10.7 wt.-%, respectively. These water yields correspond to a total recovery rate for these three experiments of 98.3 wt.-%, 102.8 wt.-% and 95.4 wt.-%. The overall mass balance can thus, especially for hydrolysis lignin pyrolysis, be closed mainly by reaction water. This conclusion is also supported by literature data and the atomic H/O-ratio of the loss which is close to 2 for hydrolysis lignin. In literature, for example, the pyrolysis of wheat straw lignin, with a purity of about 80 wt.-%, in a centrifuge reactor yielded 10 wt.-% reaction water [124]. This reaction water is predominantly believed to originate from dehydration reactions, i.e. from hydroxyl groups [119, 323].

6.2.1.3 Product composition and component yields

Char composition

Char ultimate and proximate analyses of selected samples are listed in Table 6.4. For every pyrolysis temperature and both lignins the chars obtained from the fluidized bed directly (BM) and from the secondary cyclone (C2) have been analyzed. Especially for Kraft lignin it can be observed that the char content in the bed material correlates with the amount of lignin fed into the CFB reactor (cf. also Table 4.7). But due to the high amount of quartz sand in the samples, it is difficult to obtain good analyses for the ultimate char composition. Specifically, the oxygen content, which is calculated by difference, can only be calculated with a relatively large error. Here the char's particle morphology (cf. Section 6.1) is beneficial. Kraft lignin char contains char layer fragments which are smaller than the quartz sand particle size and hydrolysis lignin char contains particles smaller and larger than the lower and upper particle size distribution boundaries of the quartz sand. From evaluation of the distributions for Kraft lignin char C2 < 71 µm and for hydrolysis lignin char C2 < 71 µm and BM > 710 µm were used (cf. Fig. 6.1). Therefore, from sieving in most cases relatively pure char samples could be obtained. For example, the secondary cyclone material from experiment 78 has an inert content of 12.8 wt.-%, which is quite close to the theoretical value of 5.3 wt.-% (assuming that the total amount of ash remains in char; $w_{ash,L}/Y_{char} \cdot 100$). For hydrolysis lignin char this theoretical value (about 50 wt.-% ash content) lies in between the inert content values of BM > 710 µm and C2 < 71 µm. Due to the better and visually perfect separation of large char particles from quartz sand it can be expected that the smaller value (30 to 40 wt.-%) is close to or equal to the hydrolysis lignin char's ash content. The influence of temperature on char composition (i.e. especially the content of hydrogen and oxygen) is further depicted in Fig. 6.13 and discussed below.

Table 6.4: Ultimate and proximated analyses for selected char samples

exp. no.	lignin type	sample	ϑ °C	mass fraction wt.-%, dry basis						
				C	H	$O^\dagger$	N	S	$I^\ddagger$	M^*
78	KL	BM	550	10.23	0.10	0.52	0.03	0.13	89.0	0.0
78	KL	C2	550	30.85	0.40	1.57	0.08	0.40	66.7	0.1
78	KL	C2 < 71 µm	550	80.02	1.86	4.07	0.22	1.05	12.8	1.7
63	KL	BM	600	6.53	0.15	0.29	0.11	0.07	92.8	0.5
63	KL	C2	600	12.31	0.35	0.55	0.13	0.15	86.5	0.8
63	KL	C2 < 71 µm	600	81.07	1.40	3.60	0.24	1.05	12.6	2.2
89	KL	BM	650	19.67	0.10	1.03	0.10	0.25	78.8	0.3
89	KL	C2	650	38.10	0.50	2.00	0.12	0.51	58.8	0.0
89	KL	C2 < 71 µm	650	80.30	1.82	4.22	0.25	1.08	12.3	1.8
64	KL	BM	700	1.59	0.02	0.03	0.09	0.03	98.2	0.1
64	KL	C2	700	3.69	0.07	0.07	0.09	0.07	96.0	0.2
64	KL	C2 < 71 µm	700	22.63	0.16	0.43	0.06	0.44	76.3	1.3
91	HL	BM	500	7.26	0.25	1.03	0.15	0.01	91.3	0.0
91	HL	C2	500	17.16	0.79	3.83	0.37	0.05	77.8	0.0
91	HL	BM > 710 µm	500	58.10	2.00	8.23	1.20	0.10	30.4	0.0
91	HL	C2 < 71 µm	500	36.70	1.70	8.20	0.80	0.10	52.5	0.0
88	HL	BM	600	6.46	0.15	0.58	0.14	0.01	92.7	0.0
88	HL	C2	600	18.68	0.50	3.33	0.43	0.12	76.9	0.0
88	HL	BM > 710 µm	600	59.30	1.40	5.32	1.30	0.10	32.6	0.9
88	HL	C2 < 71 µm	600	30.30	1.00	5.40	0.70	0.20	62.4	0.3
90	HL	BM	700	4.64	0.05	0.28	0.08	0.02	94.9	0.0
90	HL	C2	700	19.19	0.20	2.23	0.33	0.07	78.0	0.0
90	HL	BM > 710 µm	700	54.00	0.60	3.22	0.90	0.20	41.1	0.2
90	HL	C2 < 71 µm	700	29.50	0.50	3.43	0.50	0.10	66.0	0.0

†by difference; ‡inert (quartz sand and ash); *M: Moisture, as received

Figure 6.13 shows the composition of pyrolysis char in comparison to lignin feedstock and pyrolysis oil in the van Krevelen [324] diagram. The values for char have been calculated from average values of all experiments obtained at the same temperature. Due to the improved accuracy the sieved char samples BM > 710 µm and C2 < 71 µm are used. Kraft lignin has a lower O/C- and H/C-ratio than hydrolysis lignin; as carbohydrates contain more oxygen and hydrogen than lignin (cf. Table 4.4). The result is that the Kraft lignin char also has lower O/C- and H/C-ratios than hydrolysis lignin char. The trend for pyrolysis char is similar to that of pyrolysis oil. With rising pyrolysis temperature and with lower oxygen and hydrogen content in the lignin feedstock the carbonization degree increases. However, both H/C- and O/C-ratio of pyrolysis char are much lower than of pyrolysis oil. Furthermore, it can be seen that the carbonization degree of the hydrolysis lignin bed material is higher than that of the secondary cyclone material due to the longer residence time of the bed material at pyrolysis conditions. As the high content of inert sand in the bed material made the determination of the char composition difficult, for Kraft lignin only secondary cyclone material is depicted. However, it can also be expected for Kraft lignin that the carbonization degree of char contained in the bed material is higher than for the char contained in secondary cyclone material. These findings are in good agreement with the composition of Kraft lignin char (cf. also Fig. 6.13) produced in a tubular reactor at 150 to 550 °C [123]. The Kraft lignin char carbonization degree in the tubular reactor also increases with rising temperature. However, the carbonization degree of this Kraft lignin char is lower than in the circulating fluidized bed, which can be explained by a shorter residence time of 10 min and possibly a lower heating rate.

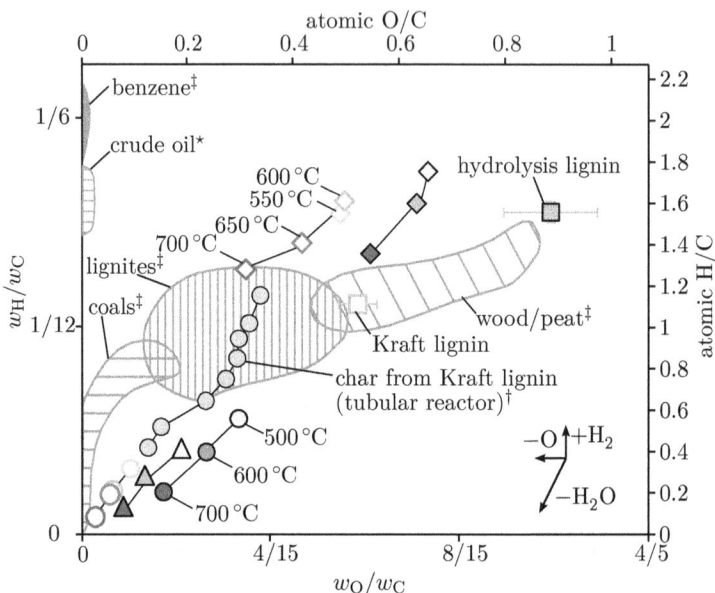

Figure 6.13: van Krevelen diagram for Kraft lignin (empty symbols) and hydrolysis lignin (filled symbols): ■ lignin, ◆ pyrolysis oil, ● secondary cyclone material, and ▲ bed material with color code: rising temperature

Literature data: [†]alkali lignin char depending on pyrolysis temperature (lignin upmost right symbol to lignin char at 550 °C lowermost left symbol) [123]; [‡][325]; [*][326]

Oil composition and component yields

The pyrolysis oil, sampled as described in Section 4.2.1, consists of both monomeric and oligomeric constituents. Approximately more than 80 % of both lignin oils are oligomeric substances. But it is quite challenging to analyze the oligomeric fraction in detail [46]. Thus, only few research is carried out on decoding of oligomeric structures (e.g. [46]). Instead, usually SEC is used for characterization; same is done here. The monomeric substances are analyzed by GC (cf. Section 4.2.2). Figure 6.14 (a) shows the two example chromatograms for side stream sample V78B: the unmodified sample as well as the silylated sample for quantification of catechols. On average 80 % of the chromatogram area was identified and quantified. The identified monomeric components are listed in Table 6.5 and tagged with a component identifier used for labeling in Figures 6.14 and 6.15 and annotation in the following text (bold numbers). Quantification is done from the chromatogram area information by conversion into yields on lignin weight basis. The corresponding procedure is explained in section 4.3.2. As described by various authors [36, 47, 48, 101, 102, 113, 114, 123, 156, 162, 185, 186] the monomeric fraction of the pyrolysis oil consists of a large number of different components.

The resulting yields of the main aromatic components are shown in the bubble plots (Figures 6.14 and 6.15), which show the relationship between temperature and retention

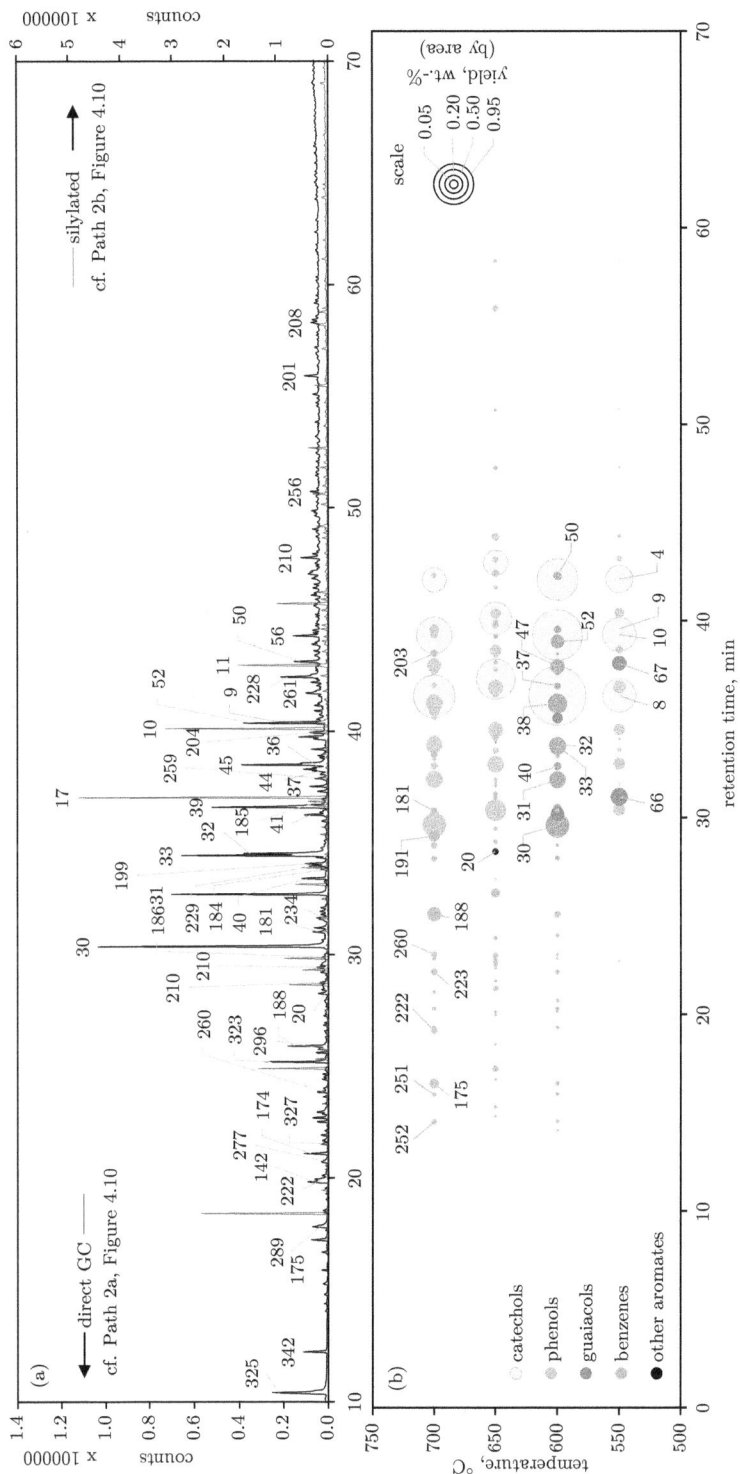

Figure 6.14: (a) Chromatograms of side stream sample V78B and (b) temperature dependent monomeric aromatics yields for Kraft lignin, component identifier cf. Table 6.5

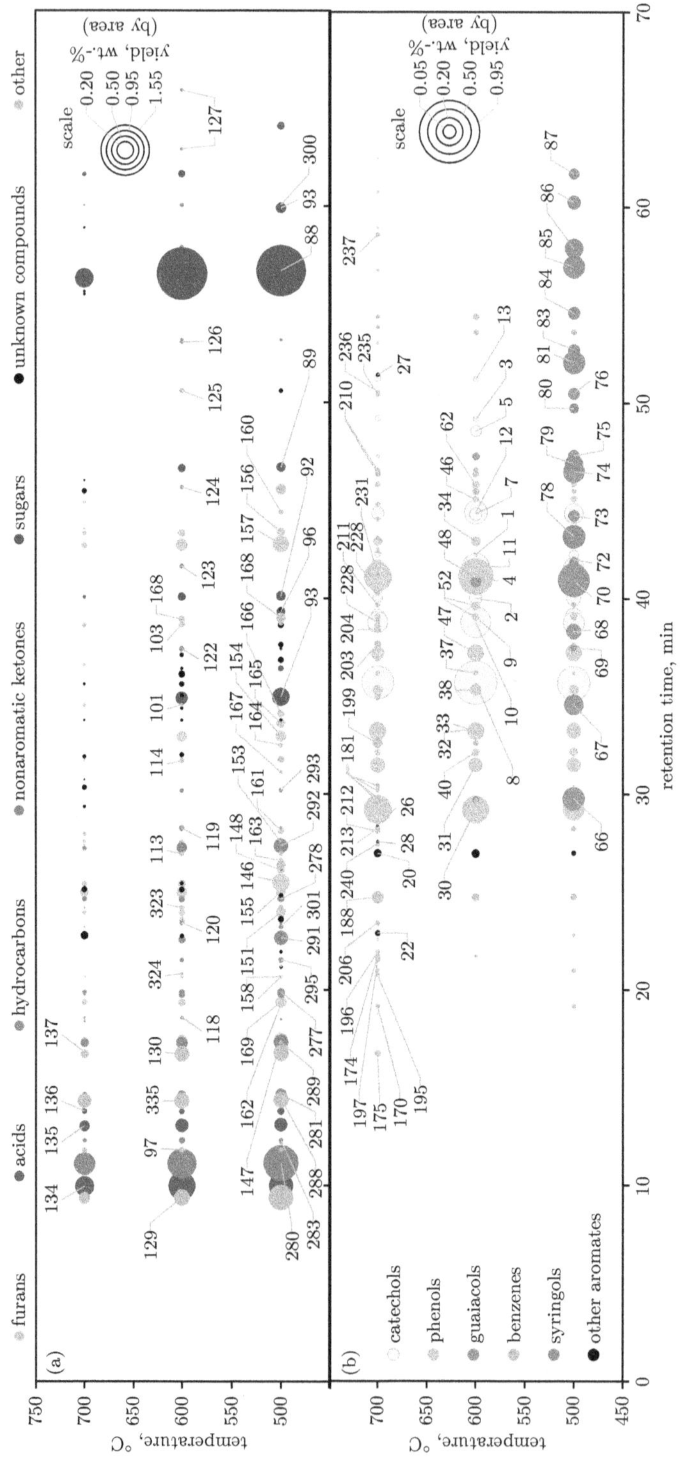

Figure 6.15: Yields of monomeric (a) non-aromatic and (b) aromatic oil compounds for hydrolysis lignin, component identifier cf. Table 6.5

time with the bubble area representing the components yield on mass basis. It should be noted that a small shift in retention time is possible due to analytical reasons. Figure 6.14 (b) displays the aromatics yields for Kraft lignin depending on temperature. It can be seen that catechols have the highest yields followed by phenols. At 600 °C the highest catechols yields are obtained with 1.9 wt.-% **8** catechol, 1.4 wt.-% **10** 4-methyl-catechol, and 1.0 wt.-% **4** dimethyl- or ethyl-benzenediol. In total that amounts to 4.7 wt.-% catechols. For phenols, which predominantly have alkyl side groups, also an optimum can be observed at 600 °C. The corresponding highest component yields are 0.3 wt.-% and 0.2 wt.-% for **30** phenol and **38** 2,4-dimethyl-phenol, respectively. Guaiacols are only found at temperatures ≤ 600 °C with a yield decreasing with increasing temperature. The yields of **66** guaiacol (0.16 wt.-%) and **67** 4-methyl-guaiacol (0.12 wt.-%) are the biggest at 550 °C. The overall yield of guaiacols at 550 °C is 0.33 wt.-%. With rising temperature the yields of benzenes rise (to 0.45 wt.-% at 700 °C), resulting in a maximum of 0.1 wt.-% **188** indene and 0.05 wt.-% **175** styrene at 700 °C. No syringols and only a few other aromates have been found for Kraft lignin (cf. also Fig. 6.16b).

Table 6.5: List of identified monomers in pyrolysis oils: component identifier and chemical classification

id.	component	cl.[†]	id.	component	cl.[†]
1	methylhydroquinone, bis(trimethylsilyl) ether	C	147	furaldehyde, 2-	F
2	silyl. hydroquinone	C	148	furan-2-one, 5-methyl-, (5H)-	F
3	silyl. catechol, propenyl-	C	151	furaldehyde, 5-methyl-2-	F
4	silyl. benzenediol, dimethyl- or ethyl-	C	153	furan-2-one, 3-methyl-, (5H)-	F
5	silyl. catechol, allyl- or propenyl-	C	154	furan-2-one, 4-methyl-(5H)-	F
7	silyl. catechol, vinyl-	C	155	butyrolactone, ?-	F
8	silyl. catechol	C	156	furanone derivative	F
9	silyl. catechol, 3-methyl-	C	157	furaldehyde, 5-(hydroxymethyl)-, 2-	F
10	silyl. catechol, 4-methyl-	C	158	ethanone, 1-(2-furanyl)-	F
11	silyl. catechol, ethyl- or dimethyl-	C	160	2(3H)-furanone, dihydro-4-hydroxy-	F
12	silyl. benzenetriol	C	161	furan-2-one, 2,5-dihydro-3,5-dimethyl-	F
13	silyl. catechol, C$_3$-	C	162	furan-2-one, 2,3-dihydro-5-methyl-	F
20	benzaldehyde, 2-hydroxy	AAl	163	furandione-2,5-, 3-methyl-	F
22	benzaldehyde	AAl	164	furandicarboxaldehyde, 2,5-	F
26	acetophenone	AK	165	lactone derivative	F
27	acetophenone, 3-hydroxy-	AK	166	isom. of furan-2-one, 2,5-dihydro-3,5-dimethyl-	F
28	acetic acid, phenyl ester	AE	167	furancarboxylic acid, methyl ester, 3-	F
30	phenol	P	168	2H-pyran-3(4H)-one, dihydro-6-methyl-	F
31	cresol, o-	P	169	furfuryl alcohol, 2-	F
32	cresol, m-	P	170	toluene, 3-ethyl-	B
33	cresol, p-	P	174	benzene, 1-propenyl-	B
34	phenol, 4-propenyl-(cis)	P	175	styrene	B
36	phenol, 2,3-dimethyl-	P	181	1H-indene, methyl-	B
37	phenol, 2,4,6-trimethyl-	P	184	1H-indene, dimethyl-	B
38	phenol, 2,4-dimethyl-	P	185	1H-indene, ethyl-	B
39	phenol, 2,5-dimethyl-	P	186	1H-indene, ethyl- or dimethyl-	B
40	phenol, 2,6-dimethyl-	P	188	indene	B
41	phenol, 2-ethyl-	P	191	1H-indene, 2,3-dihydro-dimethyl-	B
44	phenol, 3,5-dimethyl-	P	195	benzene, 1-ethenyl-4-methyl- or indane	B
45	phenol, 3-ethyl-	P	196	benzene, 1-ethenyl-2-methyl-	B
46	phenol, 3-methoxy-5-methyl-	P	197	benzene, 2-propenyl-	B
47	phenol, 4-ethyl-	P	199	naphthalene	B
48	phenol, 4-vinyl-	P	201	naphthalenol, 2-	B
50	phenol, dimethyl-ethyl-	P	203	naphthalene, 2-methyl-	B
52	phenol, ethyl-methyl-	P	204	naphthalene, 1-methyl-	B

Table 6.5: (continued)

id.	component	cl.[†]	id.	component	cl.[†]
56	poss: phenol, allyl- or propenyl-	P	206	benzene (allyl- or propenyl)	B
62	phenol, 4-propenyl-(trans)	P	208	naphthalenol, methyl-	B
66	guaiacol	G	210	benzofuran, methyl-	B
67	guaiacol, 4-methyl-	G	211	inden-1-one, 2,3-dihydro-1H-	B
68	guaiacol, 4-ethyl-	G	212	benzofuran, 7-methyl-	B
69	poss: guaiacol, 3-ethyl-	G	213	1H-indene, 2,3-dihydro-methyl-	B
70	guaiacol, 4-vinyl-	G	214	benzofuran, methyl-	B
72	guaiacol, 4-propyl-	G	215	benzene, (1-methylethenyl)-	B
73	guaiacol, 4-propenyl-(cis)	G	222	benzene, ethyl-methyl-	B
74	guaiacol, 4-propenyl-(trans)	G	223	benzene, ethyl-dimethyl-	B
75	vanillin	G	228	benzofuran, dihydro-	B
76	phenylethanone, 4-hydroxy-3-methoxy-	G	229	benzofuran, ethyl-	B
78	syringol	Sy	231	poss: biphenyl	B
79	syringol, 4-methyl-	Sy	234	$C_{10}H_{10}$ (unsat. benzene compound)	B
80	syringol, 4-ethyl-	Sy	235	fluorene	B
81	syringol, 4-vinyl-	Sy	236	fluorene, 2,4a-dihydro-	B
83	syringol, 4-propyl-	Sy	237	phenanthrene	B
84	syringol, 4-(1-propenyl)-(cis)	Sy	240	benzofuran, 2-methyl-	B
85	syringol, 4-(1-propenyl)-(trans)	Sy	243	poss: benzene, (2-methyl-1-propenyl)-	B
86	syringaldehyde	Sy	250	xylene, m-	B
87	syringyl acetone	Sy	251	xylene, o-	B
88	levoglucosan	S	252	xylene, p-	B
89	anhydro-β-D-xylofuranose, 1,5-	S	256	naphthalene, tetramethyl-	B
92	dianhydro-α-D-glucopyranose, 1,4:3,6-	S	259	1H-indene, ethyl-methyl-	B
93	anhydrosugar (unknown)	S	260	benzofuran	B
96	2,3-anhydro-d-mannosan	S	261	benzofuran, 2,3-dihydro-	B
97	pyridine	N	265	isomere of styrene	B
101	benzyl nitrile	N	277	cyclopenten-1-one, 2-methyl-2-	K
103	pyridinol, 3-	N	278	cyclopenten-1-one, 3-methyl-2-	K
113	pyran-2-one, 2H-	Py	280	acetol (Hydroxypropanone)	K
114	maltol	Py	281	butanone, 1-hydroxy-2-	K
118	decene, 1-	K	283	acetoin (hydroxy-2-butanone, 3-)	K
119	dodecene, 1-	H	288	cyclopentanone	K
120	undecene, 1-	H	289	cyclopenten-1-one, 2-	K
122	tetradecene, 1-	H	291	cyclopenten-1-one, 2-hydroxy-2-	K
123	pentadecene, 1-	H	292	cyclopenten-3-one, 2-hydroxy-1-methyl-1-	K
124	hexadecene, 1-	H	293	cylopenten-1-one, 3-ethyl-2-	K
125	nonadecene, 1-	H	295	cyclopentene-1,3-dione, 4-	K
126	unknown aliphatic compound	H	296	isom. of 2-cyclopenten-1-one, 3-methyl-	K
127	unknown alkene	H	300	heptadecanone, 2-	K
129	acetaldehyde, hydroxy-	Al	301	acetonylacetone (hexandione, 2,5-)	K
130	butyraldehyde, x-hydroxy-oxo-	Al	323	propane, 2,2',2''-[methylidyne-tris(oxy)]tris-	M
134	acetic acid	Ac	324	dimethyl sulfoxide	M
135	propionic acid	Ac	325	propane, 2,2'-[methylenebis(oxy)]bis-	M
136	2-propenoic acid	Ac	327	tetrahydrophthalic anhydrid	M
137	butyric acid	Ac	335	ethyleneglycol	Alc
142	vinylfuran	F	342	propane, 2,2'-[ethylidenebis(oxy)]bis-	M
146	furanone, 2(5H)-	F			

[†]AAl: aromatic aldehydes, Ac: acids, AE: aromatic esters, AK: aromatic ketones, Al: aldehydes (non-aromatic), Alc: alcohols (non-aromatic), B: benzenes, C: catechols, F: furans, G: guaiacols, H: hydrocarbons, K: ketones (non-aromatic), M: miscellaneous, N: N-compounds, P: phenols, Py: pyrans, S: sugars, Sy: syringols

On the contrary, the pyrolysis of hydrolysis lignin yields also a considerable amount of syringols and a larger number and yield of guaiacols (Fig. 6.15 (b)). This larger product spectrum is caused by the lignin origin (cf. Section 2.1.2: Kraft lignin from softwood is a G-lignin, whereas hydrolysis lignin from wheat straw is a GSH-lignin). With rising temperature functional groups (mainly $-OCH_3$ and $-OH$) are cleaved from the aromatic rings, thus the amount of guaiacols in the pyrolysis oil of both lignins and of syringols in hydrolysis lignin oil decrease with rising temperature. Therefore, syringols can only be found at 500 °C and guaiacols are substantially decreased in yield from 500 to 600 °C. It is, for example, believed that catechols form from primary homolysis of guaiacols and syringols [327]. The main guaiacols and syringols at 500 °C are **70** 4-vinyl-guaiacol (0.26 wt.-%), **66** guaiacol (0.13 wt.-%), **78** syringol (0.14 wt.-%), **81** 4-vinyl- and **85** 4-(1-propenyl)-(cis)-syringol (both 0.13 wt.-%). Above 600 °C the yields of catechols for both lignins and phenols for Kraft lignin decrease. Thus at these elevated temperatures dehydration reactions are favored. The main catechols found in hydrolysis lignin oil are **8** catechol and **10** 4-methyl-catechol with a yield of 0.46 wt.-% and 0.23 wt.-%, respectively. Furthermore, the demethoxylation and scission of R−OH bonds are also the reason for the increase of benzenes yield with rising pyrolysis temperature. Next to **188** indene and **175** styrene (0.04 wt.-% and 0.01 wt.-%) also **199** naphthalene (0.03 wt.-%) is a product of hydrolysis lignin pyrolysis at 700 °C. The yield of phenols is highest at 600 °C. The main phenolic products at 600 °C are **48** 4-vinyl-phenol and **30** phenol with yields of 0.31 wt.-% and 0.18 wt.-%, respectively. It can be observed that the Kraft lignin oil contains less different catechols and guaiacols than the hydrolysis lignin oil.

Aromatic monomers found in pyrolysis oils originate from the cleavage of the interconnections between the aromatic components contained in the lignin structure but not from carbohydrates. Therefore, it is useful to compare also the yields of monomeric aromatics on the basis of pure lignin (Y_i/w_L), shown in Fig. 6.16. As the hydrolysis lignin has a purity of about 49 wt.-% the yield based on pure lignin is about twice as high compared to feed lignin basis. Fig. 6.16a depicts the values for catechols, phenols, and benzenes. Although the total yield on feed basis of phenols is higher for Kraft lignin (at $\vartheta \geq 600$ °C), the yield is higher on pure lignin basis for hydrolysis lignin. Furthermore, it can be seen in Fig. 6.16b that at 500 °C and 550 °C the yields of guaiacols, syringols, and phenols for hydrolysis lignin and guaiacols and phenols for Kraft lignin have the same magnitude, respectively. However, from Kraft lignin fewer phenols and guaiacols are formed. Either the presence of carbohydrates or the different lignin structure, containing e.g. syringols is thus beneficial for the formation of phenols, guaiacols, and syringols. De Wild at al. [29] pyrolyzed wheat straw lignins of high purity (≥ 92.7 wt.-%) in a fluidized bed at 500 °C. The yields of guaiacols, syringols, alkylphenols and catechols were found to be 2 wt.-%, 0.8 to 1 wt.-%, 0.4 to 0.5 wt.-% and 0.4 to 0.5 wt.-%. Compared to the yields on pure lignin basis in Figure 6.16 – with exception of guaiacols – a higher yield was obtained in CFB pyrolysis. The higher yield could be attributed to the presumably lower vapor residence time in the CFB system, a different lignin structure, or the beneficial influence of carbohydrates contained in the hydrolysis lignin.

The number of chemical compounds in the pyrolysis oil is much higher for hydrolysis lignin compared to Kraft lignin due to the components originating from the larger carbohydrate content of the hydrolysis lignin. While the oil components of carbohydrate origin in Kraft lignin are of negligible number and yield – with one exception, discussed below – they have to be considered for hydrolysis lignin. Figures 6.15 (a) and 6.17 show the yields the compounds from carbohydrate origin in dependence on pyrolysis temperature. In Figure 6.15 (a) the quantified components are roughly grouped in furans, acids, hydrocarbons

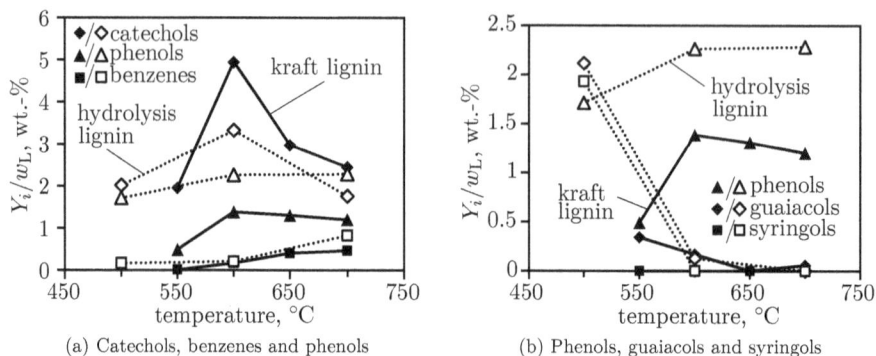

(a) Catechols, benzenes and phenols

(b) Phenols, guaiacols and syringols

Figure 6.16: Yields (based on pure lignin, i.e. scaled by lignin content w_L) of aromatic groups for Kraft and hydrolysis lignin

(alkenes and alkanes), non-aromatic ketones, sugars, unknown components (which could not be identified by the available in-house and NIST standards), and others (pyrans, nitrogen-containing compounds, non-aromatic esters, alcohols and aldehydes). In Figure 6.17 the yields of sugars, heterocyclic (furans and pyrans) and non-aromatic compounds (acids, hydrocarbons (alkenes and alkanes) and non-aromatic ketones, esters, alcohols, and aldehydes) are depicted. As mentioned above, the number and yield of non-aromatic components in Kraft lignin oil are very small except for 1-propanol, which is the component increasing the non-aromatic compounds yield in Fig. 6.17 (a) from 0 to 1.87 wt.-%. The alcohol 1-propanol might form from the alcoholic side chains of the phenylpropane (C_9) subunits coumaryl and coniferyl alcohol. For hydrolysis lignin, on the other hand, a steep yield decrease of sugars, non-aromatic and heterocyclic compounds can be observed with increasing temperature: 1.94 to 0.24 wt.-%, 2.73 to 1.02 wt.-% and 1.29 to 0.17 wt.-%, respectively. The component in hydrolysis lignin pyrolysis oil with the highest yield is **88** levoglucosan with 1.62 wt.-% at 600 °C. Of the non-aromatic compounds in Fig. 6.17b ketones, acids, and aldehydes have the highest yields. The main components are **280** acetol, **289** 2-cyclopenten-1-one, **292** 2-hydroxy-1-methyl-1-cyclopenten-3-one, **134** acetic acid, **135** propionic acid, and **129** hydroxy-acetaldehyde. The corresponding maximum yields (at 500 °C) are 0.77 wt.-%, 0.13 wt.-%, 0.13 wt.-%, 0.36 wt.-%, 0.1 wt.-% and 0.38 wt.-%, respectively. The yields of hydrocarbons and acids are highest at 600 °C. The main alcohol in hydrolysis lignin oil is **335** ethyleneglycol with constantly 0.15 wt.-%. The heterocyclic compounds consist up to about 85 % of furans, the remaining 15 % are pyrans. **146** 2-(5H)-furanone and **157** 5-(hydroxymethyl)-,2-furaldehyde are most prevalent furans with yields of 0.19 wt.-% and 0.17 wt.-%, respectively.

Applicability of pyrolysis oil

Figure 6.13 shows the composition of pyrolysis oil in comparison to lignin and char in the van Krevelen diagram. The O/C-ratio of pyrolysis oil depends on the feedstock composition and pyrolysis temperature (Figure 6.13). It can be seen that hydrolysis lignin oil has a higher O/C-ratio than pyrolysis oil derived from Kraft lignin. During pyrolysis of hydrolysis lignin more deoxygenation reactions proceed, resulting in a higher decrease of O/C-ratio, but as more oxygen is contained in hydrolysis lignin than in Kraft lignin the final O/C-ratio in hydrolysis lignin oil is higher. An increase in pyrolysis temperature decreases both O/C- and H/C-ratio due to intensified cracking conditions. This composition is of importance for applicability as besides the high water content low aliphatic content,

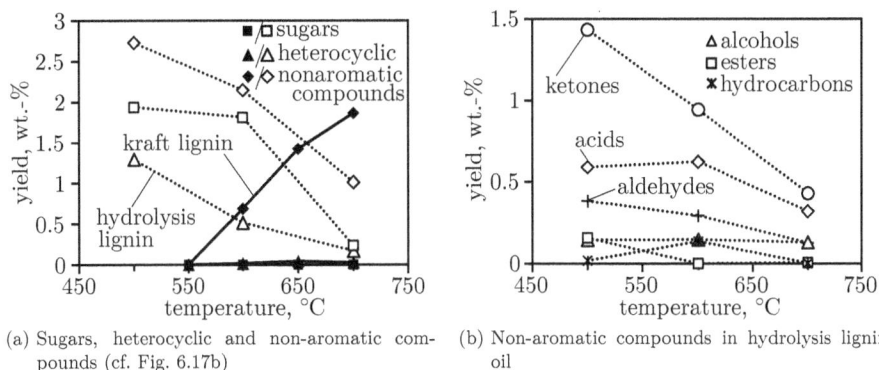

(a) Sugars, heterocyclic and non-aromatic compounds (cf. Fig. 6.17b)

(b) Non-aromatic compounds in hydrolysis lignin oil

Figure 6.17: Yields of non-aromatic groups for Kraft and hydrolysis lignin

high viscosity, and high molar weight of pyrolysis oil, it's high oxygen content is the major impediment to technical application [328]. Especially the high oxygen content affects the bio-oil stability negatively, due to the tendency of pyrolysis oil separation into two phases as well as pyrolysis oil oxidization at high oxygen content (stability issue). In principle biomass pyrolysis oil can be utilized as co-feedstock of conventional crude oil refineries [329, 330]. Nevertheless, the oxygen content is significantly higher in the lignin bio-oil than in crude oil (cf. Figure 6.13) and thus stabilization by deoxygenation of oxygenated compounds such as acids, aldehydes, esters, phenolics, furanics and oxygenated oligomers [331, 332] is desirable. Unfortunately, the high diversity of bio-oil composition is a major problem in accomplishing this aim [333]. Therefore, the research activity in upgrading techniques has increased recently [334–336]. Three main catalytic upgrading techniques have been identified and are currently investigated for oxygen removal of pyrolysis oil [337]. These techniques are hydrodeoxygenation (HDO) [338–341], hydrogenation, and condensation reactions [342–344]. The condensation reactions proceed without addition of hydrogen, whereas hydrodeoxygenation and hydrogenation need the supplement of hydrogen. Hydrodeoxygenation and hydrogenation are expensive due to the high H_2 demand. Furthermore, the ecological benefit of the bio-oil production route is reduced as today about 96 % of hydrogen is produced from fossil fuels [345]. This situation may change in the future, when more hydrogen is produced from renewable sources [346]. The condensation reaction pathways, i.e. ketonization, aldol condensation, and acid-catalyzed esterification are accompanied by the release of water and if so CO_2. Hydrodeoxygenation (HDO) releases oxygen as water by means of hydrogen addition, whereas hydrogenation saturates double bonds in organic acids (ketones), aldehydes and alkenes. Therefore, hydrodeoxygenation, hydrogenation and/or condensation might be promising catalytic upgrading steps for pyrolysis oil.

Gas composition and component yields

Figure 6.18 shows the average gas component yields for the two lignins in dependence of the temperature. For Kraft lignin, the hydrogen content was not measured for all experiments. Therefore, the hydrogen yield is not depicted for Kraft lignin, but at 650 °C a H_2 yield in the range of 0.1 to 0.9 wt.-% was determined. At 700 °C no gas sample bag was taken, thus only the integral hydrocarbon yield is given. The gas yield of hydrolysis lignin is higher than that of Kraft lignin. It can be observed that the difference in total yield comes mainly from a CO yield that is about 35 % higher at 600 and 700 °C for hydrolysis lignin. The higher CO yield can be attributed to the higher cellulose content of hydrolysis lignin. Cellulose has a higher carbonyl content than the other two main biomass components and CO is mainly formed from cracking of carbonyl COC and carboxyl C=O groups [102, 119]. While the CO yield does significantly increase from about 6 to 25 wt.-% for hydrolysis lignin and from 8 to 15 wt.-% for Kraft lignin, the CO_2 yield is not affected to a large extent for either lignin in the considered temperature range, as no clear trend can be observed. The CO_2 fluctuates in a range of 5.5 to 10.4 wt.-% for Kraft and 6.9 to 8.4 wt.-% for hydrolysis lignin. CO_2 is formed mainly from cracking and reforming of carboxyl C=O and COOH functional groups [102]. Other main permanent gas components are the hydrocarbons methane, ethene, ethane, propene, propane, and a low amount of acetylene, which are assumed to form from mainly aliphatic and alicyclic (lignin) moieties [119]. But CH_4 is the largest hydrocarbon fraction. It is the main degradation product of methoxy–O–CH_3 and methyl side groups which are more frequent in lignin (compared to cellulose and hemicellulose) [347]. For both lignins, the hydrocarbons yield increases with rising temperature from 2.3 to 7.2 wt.-% and from 2.0 to 7.3 wt.-% for Kraft and hydrolysis lignin, respectively. This yield conformity, at first sight, indicates that the evolution of hydrocarbons is not majorly affected by lignin composition in this work. But in contrast to these findings, Qu et al. [102] stated that most CH_4 is generated in lignin pyrolysis when compared to xylan and cellulose. Qu et al. [102] found the CH_4 gas fraction in lignin pyrolysis gas to be up to four times higher compared to cellulose and hemicellulose. Furthermore, it was found that the gas composition is affected by different biomass composition and therefore a biomass (mixture of cellulose, hemicellulose, and lignin), when pyrolyzed, might produce a different gas than the sum of its components [225].

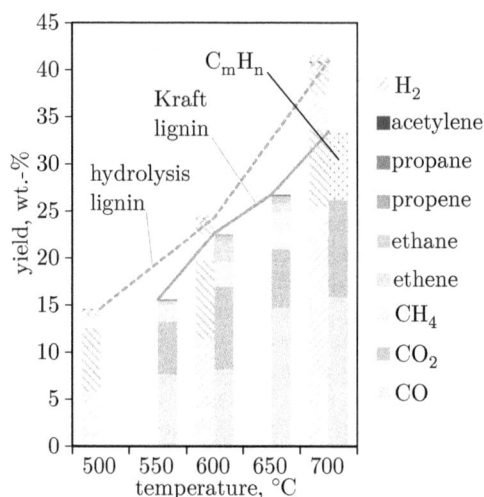

Figure 6.18: Yields of pyrolysis gas components for Kraft lignin (solid) and hydrolysis lignin (dashed)

Additionally, the lignin composition could be a reason for the similar hydrocarbons evolution for both lignins. Although the hydrolysis lignin has a much lower purity, straw lignin has – due to its structure, which contains more syringols – a higher content of methoxy side groups. The content is in the magnitude of 1.2 MeO/C_9 for lignin from grasses and 0.9 MeO/C_9 for softwood lignin [348]. Thus, it can be expected that partly more CH_4 is formed from the lignin in hydrolysis lignin counterbalancing the lower lignin purity. With rising pyrolysis temperature for hydrolysis lignin the H_2 yield increases from below detection limit to 0.7 wt.-% at 700 °C, which has the same magnitude as the determined range of Kraft lignin H_2 yield at 650 °C. It is believed that at lower temperatures, analogously to coal pyrolysis, aliphatic and alicyclic hydrogen is released in form of cracked light hydrocarbons, whereas at T>500 °C molecular H_2 is released due to condensation and rearrangement of aromatic rings [119]. This finding is in good agreement with the decreasing H/C-ratio (increasing carbonization) of the produced char (cf. Figure 6.13).

The effect of pyrolysis temperature on product gas composition is also shown in Figure 6.19. The volume fraction of the main components CO, CO_2, and C_mH_n is depicted as determined by online measurement. The hydrogen volume fraction can be up to 20 vol.-% at 650 °C for Kraft and 700 °C for hydrolysis lignin, respectively. But, it is not included in Figure 6.19 as no data is available for Kraft lignin at 550, 600, and 700 °C. For Kraft lignin the C_mH_n yield rises with increasing pyrolysis temperature as more oil molecules are cracked. Therefore, the volume fraction of C_mH_n rises by about 30 %. The increase of C_mH_n yield has the same magnitude as the gas yield so that the volume fraction does not change significantly. The same is true for CO during pyrolysis of Kraft lignin, whereas the volume fraction of CO does increase for hydrolysis lignin due to its higher carbohydrate content. The CO_2 volume fraction does decrease for both lignins. The decrease in the considered temperature range for Kraft lignin is 25.8 to 18.8 vol.-% and for hydrolysis lignin 33.6 to 12.9 vol.-%.

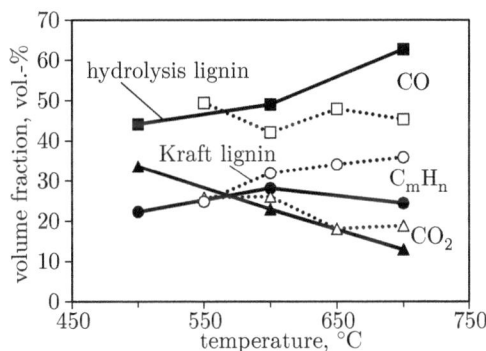

Figure 6.19: Gas volume fractions of CO, CO_2, and C_mH_n for Kraft (empty symbols) and hydrolysis lignin (filled symbols): ■ CO, ▲ CO_2, and ● C_mH_n

6.2.2 Influence of char holdup

In this section, the influence of the duration of the pyrolysis experiment is discussed. The SPE pyrolysis plant works semi-continuously, i.e. lignin is fed into and the gaseous pyrolysis products are removed from the riser reactor continuously, while the char partly remains in the CFB system (cf. Section 6.1). Therefore, char accumulates with rising amount of lignin fed into the system in the bed material. This accumulation has an influence on the pyrolysis product distribution. As the char holdup in the CFB reactor cannot be measured in-situ, but only be determined at the end of each experiment, the accumulated lignin feed is used for illustration of the phenomenon.

Figure 6.20a shows the char, gas, and total liquid (oil and water) yield dependency on accumulated lignin feed for three selected experiments at 650 °C. The char yield is determined after each experiment giving the points in Figure 6.20a. The fitted char yield from Figure 6.9a (converted to accumulated lignin feed) gives the dashed line, which fits the data points of the three experiments well. The line represents the trend for the char yield, which increases with the rise of accumulated lignin feed. The increase is 95.7 % from 12.3 wt.-% within the range of 1 to 5 kg. The gas yield shows an average increase from 27.9 to 36.7 wt.-% within the same range. Subtracting the measured gas yield and the obtained fit for the char yield (cf. Fig. 6.20a) from 100 wt.-% gives the oil yield shown in the figure. As both gas and char yield increase with accumulated lignin feed, the oil yield decreases by in average 58 % from 59.8 wt.-% within this range.

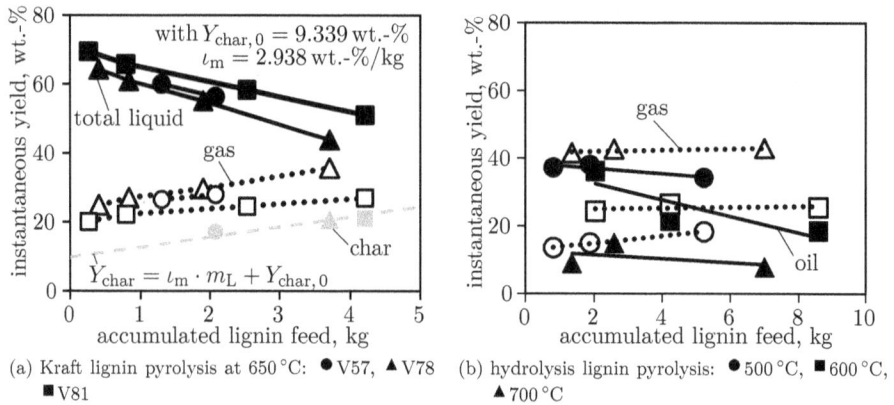

(a) Kraft lignin pyrolysis at 650 °C: ● V57, ▲ V78, ■ V81

(b) hydrolysis lignin pyrolysis: ● 500 °C, ■ 600 °C, ▲ 700 °C

Figure 6.20: Yields (instantaneous) of oil (filled symbols), gas (empty symbols), and char (gray symbols, Kraft lignin only) depending on accumulated lignin feed

Figure 6.20b shows a similar picture for hydrolysis lignin. Depicted are the yields of oil and gas for three experiments at 500, 600 and 700 °C. Again the gas yield increases, whereas the oil yield decreases with rising amount of lignin fed into the reactor. But the absolute change is considerably lower for hydrolysis lignin. In the range of 1 to 5 kg, the gas yield increases by 0.5 to 5 wt.-% and 6.5 to 12.6 wt.-% and the oil yield decreases by 3.1 to 9.6 wt.-% and 18.2 to 24.3 wt.-% for hydrolysis and Kraft lignin, respectively. Although the oil and gas yields in Fig. 6.20b underlie the same trend as for Kraft lignin – i.e. a higher decrease of oil yield than increase of gas yield with rising amount of lignin fed into the reactor – no final conclusion can be made on char yield here. Due to the lower recovery rate X_{RR} of ~ 70 wt.-% of hydrolysis lignin pyrolysis it is not possible to calculate the feeding time dependent char yield by difference. Thus, more experiments with different amounts of lignin feeding could be carried out to shed light on this issue. Furthermore, no trend can be seen for the temperature dependency, i.e. different slope of yield with changing amount of hydrolysis lignin fed and temperature. Only the increase of gas yield and decrease of oil yield with rising temperature can be observed.

During pyrolysis the accumulation of char in the bed material influences the heat transfer in the fluidized bed as well as the mineral matter concentration in the bed. The solids volume fraction and thus the gas residence time are not changed significantly with the mass of bed material in the riser increasing over time on experiment. Furthermore, it is unlikely that the heat transfer does change to an extent, which influences the yields of gas and oil significantly. Partly because of the already very good heat transfer. Nevertheless, a heat transfer impact can not be entirely excluded. To evaluate the influence of mineral matter, to answer the question why the impact is bigger for Kraft than for hydrolysis lignin, and to gain further insight, it is useful to review the inorganic species in the bed material at higher level of detail.

Table 6.6 lists accumulation factors $\chi_{i,j} = w_{i,j}/w_{i,L}$ for the measured inorganic species i in char and oil samples j in relation to the reference value $100/Y_j$. For interpretation: if $100/Y_j = \chi_{i,j}$ the inorganic species i is split proportionally to the yield of sample j, if $\chi_{i,j}$ is smaller or larger, the species is depleted or enriched in the sample, respectively. It should be mentioned at this point that the very large values for aluminum and iron most likely originate from the quartz sand composition (cf. Table 4.6). Comparing $\chi_{i,char}$ for hydrolysis lignin pyrolysis (V91, V87, V88, V90) at pyrolysis temperature ϑ ranging from 500 to 700 °C and overall lignin feed m_L ranging from 4 to almost 10 kg, shows that both parameters are insignificant in regard to $\chi_{i,j}$. Monomers extracted from the scrubber sump by methylene chloride (DCM) show negligible inorganics content, whereas the hydrolysis lignin pyrolysis tar contains mineral matter species to an amount close to or even larger than the reference value. Although the yield depending accumulation factor $\chi_{i,j}$ does not change significantly with the overall lignin feed mass m_L, the absolute mass in the riser and concentration based on riser volume does.

In Section 6.1 it was shown that Kraft lignin does spread and mainly react on the char coated bed material surface (which increases with increasing m_L), while hydrolysis lignin char particles make up a large quantity of the bed material or stick to the quartz sand surface. Therefore, with ongoing experiment, i.e. increase of char accumulation, more catalytically active inorganics become accessible to lignin particles fed into the reactor. Furthermore, it is notable that the content of Na is larger in Kraft and Ca is larger in hydrolysis lignin (cf. Table 4.4). Additionally, potassium is present in both lignins in considerably high content. These three alkaline and alkaline earth metals are known to be catalytically very important ash substances [164, 170, 171, 178–183]. Hence, the observed yield developments can mostly be explained by the rising amount of mineral matter in the bed. The here found influence on the yields of oil and char have also been shown by various researchers [164–168]. Likewise, an increase of gas yield was identified previously for some inorganic species. An example is Na^+ [178]. In detail, the sulphoxide or sulphone in Kraft lignin is known to promote lignin depolymerization [42]. Moreover, potassium has the potential to even almost double the char yield [171]. Likewise, sodium cations present in lignin increase char yield [178, 179]. Therefore, the more distinct trends for Kraft lignin can be explained by the different mineral matter content in the two investigated lignins.

Table 6.6: Inorganics accumulation in comparison to reference value $100/Y_j$

value	unit \ exp.	char (BM + C2)					DCM[††‡]	tar[†]
		V91	V87	V88	V90	V81	V87	V87
lignin	–	HL	HL	HL	HL	KL	HL	HL
ϑ	°C	500	600	600	700	650	600	600
m_L	kg	6.343	4.309	9.128	7.344	4.363	4.309	4.309
$100/Y_j$	–	4.7	4.7	5.4	5.4	4.8	40.5	13.7
Accum. factor for inorg. species i in sample j: $\chi_{i,j} = w_{i,j}/w_{i,L}$								
Fe	kg/kg	3.0	2.9	3.1	3.7	7.8	< 0.1	9.0
Mn	kg/kg	2.3	2.0	2.4	2.0	0.0	0	7.6
Zn	kg/kg	2.2	1.4	1.6	1.2	0.0	0.1	40.0
Al	kg/kg	9.7	8.2	5.8	18.6	18.5	< 0.2	8.5
Na	kg/kg	0	0	2.1	0	2.8	< 0.9	3.2
K	kg/kg	0	0	0	0	1.0	–	0
Ca	kg/kg	2.5	1.3	2.6	2.8	2.4	< 0.1	8.0
Mg	kg/kg	1.4	1.7	2.4	0.0	1.8	< 0.2	7.3
Cu	kg/kg	2.3	1.6	1.8	1.9	0	0	58.6
Ni	kg/kg	4.6	4.7	4.4	4.6	0	0	23.7
S	kg/kg	–	–	–	–	1.5	< 0.5	0

[†] obtained from scrubber, [‡] monomers extracted from scrubber sump with methylene chloride (DCM)

Oil composition and component yields

As the oil yield is affected negatively by the mineral matter contained in the fed lignin it is self-evident that also the oil composition is influenced by the accumulating pyrolysis char. Figure 6.21a depicts – in the sequence of accumulated lignin feed, labeled from A to D – the molar mass distribution of the four oil samples of experiment V81 (Kraft lignin at 650 °C). The oil samples molar mass distribution is shifted to smaller molecular sizes as the char in the reactor catalyzes diverse cleavage reactions. The oligomeric substances are cracked to form more monomers and consequently gaseous products. For further investigation on the impact of catalytically active inorganic char components on the pyrolysis process, it is further interesting to look into the development of the monomeric degradation product yields.

Although the overall oil yield decreases, the yield of many monomeric substances increases. An example are aromatics in Figure 6.21b, which rise for both lignins and have roughly the same magnitude, despite the much lower content of lignin in the hydrolysis lignin (cf. Table 4.3). The same magnitude can be explained by the positive effect of hydrocarbons on aromatics yield.

Figures 6.21c and 6.21d show the yields of catechols, phenols, and benzenes for the two lignins. The mineral matter in Kraft lignin leads to an increase of phenols and benzenes yield, whereas the catechols yield is decreased. For hydrolysis lignin also phenols and catechols are (slightly) increased, whereas the benzenes yield stays constant. The magnitude is considerably lower for hydrolysis lignin pyrolysis. The reverse catechols trends (cf. also the yields of the three main catecholic products (catechol, -methyl and -ethyl or dimethyl) depicted in Figures 6.22a and 6.22b) might be explained by the content of sodium and potassium, which is higher in Kraft lignin favoring dehydration reactions and C−O scission [179, 180, 182]. Furthermore, as the yield of phenols increases for both lignins, it can be concluded that the inorganics promote demethoxylation (cleavage of $R-OCH_3$), decarboxylation and scission of $C-C$ bonds, which is supported by the increasing yield of phenols in the presence of potassium and sodium [171, 179, 180].

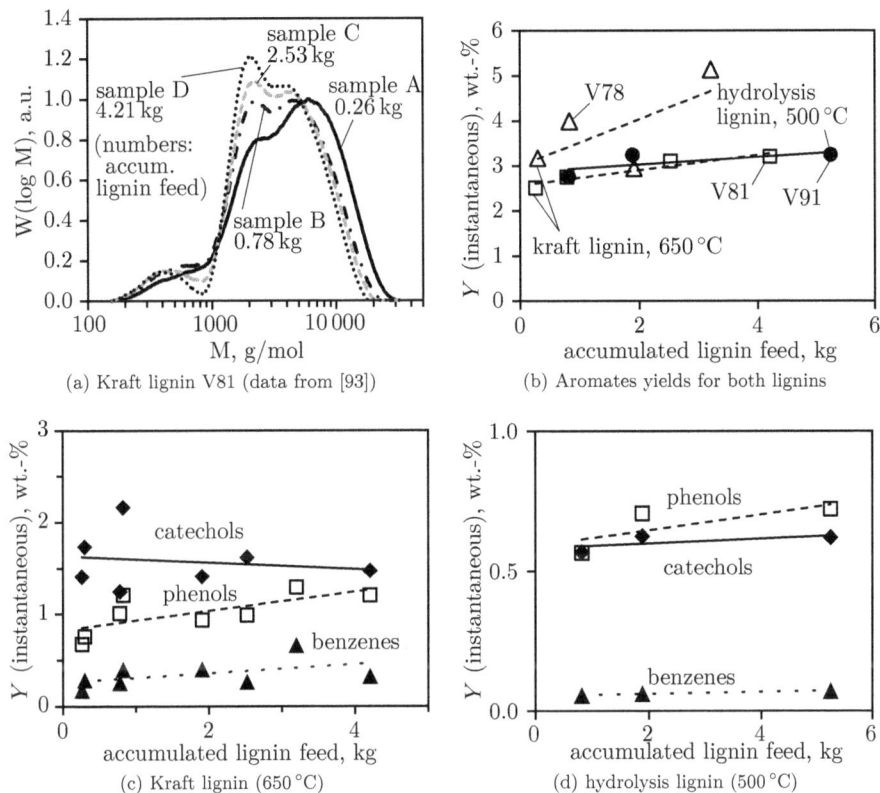

(a) Kraft lignin V81 (data from [93])

(b) Aromates yields for both lignins

(c) Kraft lignin (650 °C)

(d) hydrolysis lignin (500 °C)

Figure 6.21: Influence of char on molecular weight distribution (a) and instantaneous yields of different pyrolysis oil monomer groups (b)-(d)

The increase of cresols in Figures 6.22c and 6.22d for both lignins, on the other hand, shows that demethylation (cleavage of $R-CH_3$) is rather suppressed or at least not favored to the degree of e.g. demethoxylation or dehydration. After all, sodium is known to do just that in lignocellulosic biomass pyrolysis [179, 180]. Also, the yields of the most prevalent guaiacols and syringols in hydrolysis lignin pyrolysis, depicted in Figures 6.22e and 6.22f might be explained by the catalysis induced by the inorganics contained in the lignin/char, as potassium is known to increase guaiacol and syringol yield [171]. Although some reaction pathways are promoted or favored, whereas others are to some extent suppressed, all the aforementioned reaction pathways proceed, as the yield increase of benzenes (Fig. 6.21c) or the overall decrease of oil yield suggest.

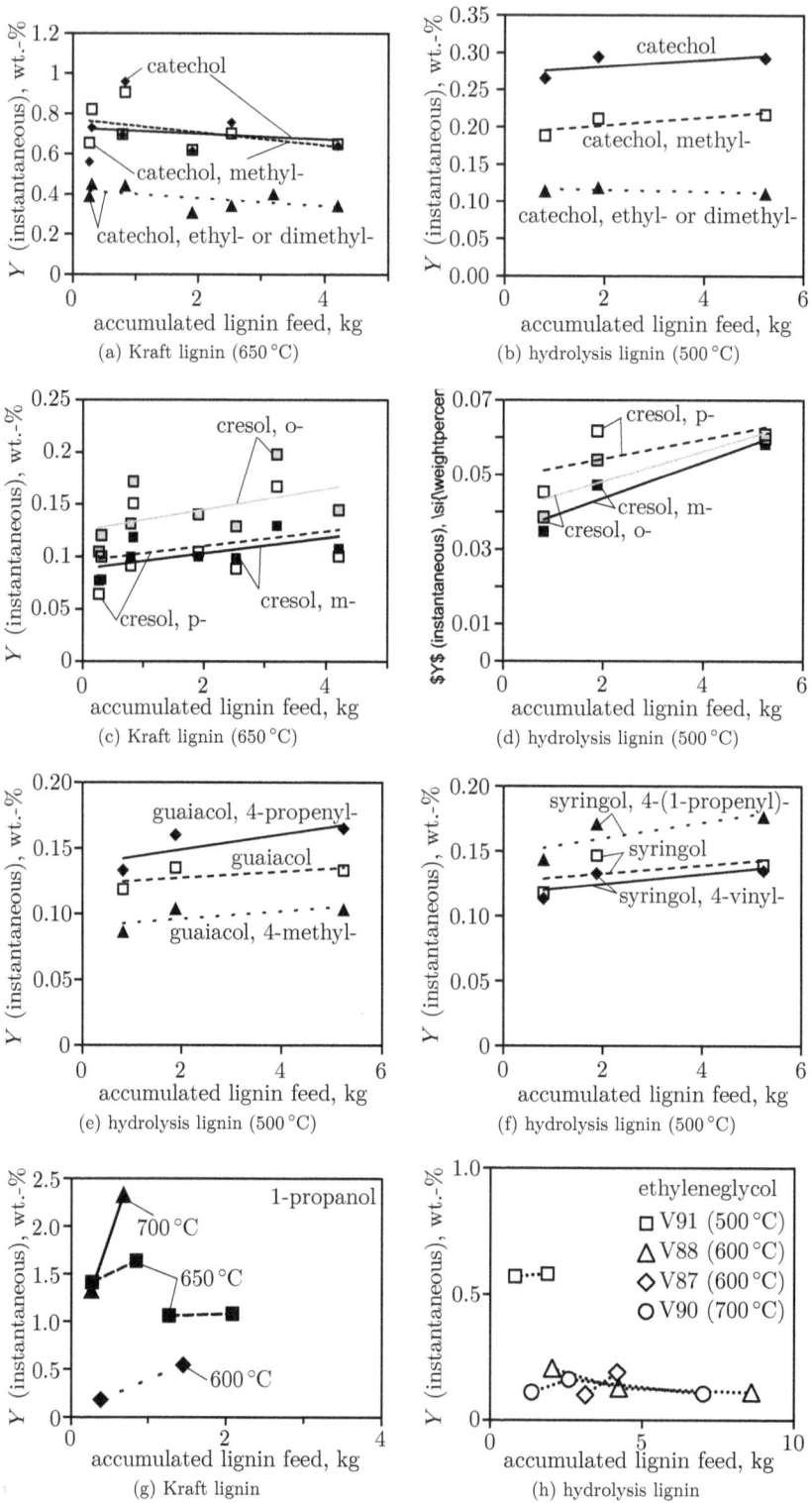

Figure 6.22: Influence of char on aromatic and non-aromatic alcohol monomer yields (instantaneous)

Also the yields of monomeric non-aromatic oil components are affected by the influence of mineral matter in char. An example is the yield development of the main alcoholic components found in the lignin oils. For Kraft lignin the yield of 1-propanol (cf. Fig. 6.22g) increases with rising temperature but also an increase with increasing amount of fed lignin can be observed. This increase might be attributed to sodium in $-COONa$ and $-CONa$ side groups, which have been found to catalytically increase the yields of alcohols [181]. Ethylene glycol (cf. Fig. 6.22h) – the main alcohol found in hydrolysis lignin – in contrary has the highest yield at the lowest temperature of 500 °C. No distinct influence of char can be observed in this case, which indicates – as both lignins contain considerable amounts of sodium – that not only the catalytic effect of sodium is important.

Gas composition and component yields

Figure 6.23 shows the gas yield and volume fraction evolution with rising accumulation of lignin feed for both lignins. For Kraft lignin, three experiments at 650 °C are compared. The yields of CO and C_mH_n rise constantly, whereas the yield of CO_2 rises faster at lower accumulated Kraft lignin feed (cf. Figure 6.23a). This yield development results in a CO volume fraction which decreases and a CO_2 volume fraction which increases before becoming almost constant. The volume fraction of C_mH_n is not significantly affected (cf. Figure 6.23c).

Figures 6.23b and 6.23d show the yields and volume fractions of the main gas components for hydrolysis lignin at 500, 600, and 700 °C. It can be seen that the yield of CO is affected by temperature the most. The yield increases from around 5.5 to over 25 wt.-% at rising temperature. The same trend can be observed for C_mH_n which rises from around 2 to over 7 wt.-% with the increase of 200 °C pyrolysis temperature. The CO_2 yield, on the other hand, is constant in the considered temperature range. Therefore, the volume fraction of CO and C_mH_n increase whereas the volume fraction of CO_2 decreases with rising temperature. With increasing accumulated lignin feed the CO yield increases at 500 °C and stays constant at 600 and 700 °C. The yield of C_mH_n increases slightly, whereas again the CO_2 yield rises at lower accumulated lignin feed before becoming constant. The volume fraction of CO decreases and of C_mH_n increases. The CO_2 volume fraction decreases at 500 °C and increases at higher temperature.

Generally, it can be concluded that the yield and volume fraction trends for both lignins go into the same direction, but the extent is smaller for hydrolysis lignin pyrolysis. Especially potassium catalyzes pyrolytic reactions, promoting the formation of CO_2 and CO [170], which is in good agreement with the yield developments found in this work. Also, sodium cations contained in lignin are known to influence the gas yield and composition [178]. As the sodium content in Kraft lignin is considerably higher than in hydrolysis lignin, this might also explain the more distinct trend for Kraft lignin gas evolution. Considered as one impact on the product distribution is that Na suppresses CO formation [179, 180]. That Na suppresses the CO formation seems to be in contradiction with the above findings for gas component yield, but regarding the development of CO volume fraction, the above findings are in good agreement with the literature [179, 180]. The main reasons for gas evolution in terms of accumulated lignin feed can thus be explained by the composition and accumulation of mineral matter in the bed material.

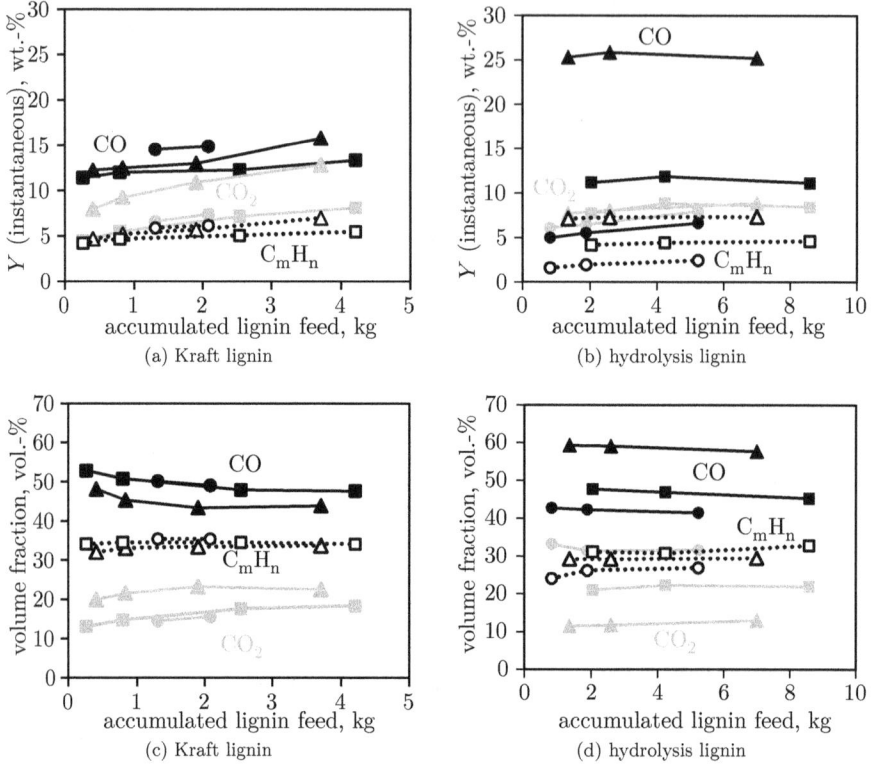

Figure 6.23: Influence of char on gas yields (instantaneous) and composition for Kraft lignin at 650 °C (experiments ● V57, ▲ V78, and ■ V81) and hydrolysis lignin at ● 500, ■ 600 and ▲ 700 °C

7 Simulation results

The discussion of the simulation results will commence with detailed sensitivity analyses about the discretization of the reactor volume into elements, the vapor residence time and biomass composition. These analyses are done with the dimensions and parameters of the CFB reactor setup used in this work. Then, the granulation behavior of the model is compared with the experimental findings. The different combinations of feedstock and pyrolysis reactor systems reported in the literature are compared to the model results. For these simulations, the reaction scheme and all parameter values of Miller and Bellan [221] are applied and coupled with the reactor model. With the obtained results the model accuracy is discussed, especially for lignin but also for general biomass pyrolysis. Subsequently, a kinetic improvement for lignin, retrieved from fitting the model curves to the experimental data provided in this work, is proposed. The resulting parameters are used for revalidation and comparison of model outcome with the different experimental combinations of feedstock and pyrolysis reactors. The chapter is completed by flowsheet modeling of an integrated pyrolysis-combustion system, which utilizes the by-products' energy. As basis, this work's CFB reactor setup with Kraft lignin feed is applied to derive an energetically feasible operation regime, while obtaining a high oil yield.

7.1 Sensitivity analyses

To investigate the influence of important model parameters sensitivity analyses are carried out. Firstly, the number of axial discretization elements is selected for a reasonable compromise between acceptable accuracy and short calculation time. Secondly, the influence of vapor residence time is shown indirectly by the modulation of the adjustable superficial gas velocity. Furthermore, the influence of biomass composition in terms of cellulose, hemicellulose, and lignin on product distribution is discussed.

7.1.1 Discretization

As criterion for evaluation of model accuracy the relative error of the model obtained total product yield compared to one is assessed in dependence on the number of discretization elements:

$$E_{\mathrm{rel}} = \frac{[Y_{\mathrm{char}} + Y_{\mathrm{oil}} + Y_{\mathrm{gas}}]_{\mathrm{model}} - 1}{1} \cdot 100\,\% \quad . \tag{7.1}$$

Table 7.1 shows, how the relative error E_{rel} converges to zero with rising number of discretization elements. At $n_{\mathrm{total}} = 2000$ a good accuracy with an error below $0.1\,\%$ is reached. Unfortunately, the calculation time on a custom personal computer can be in the order of hours for the high number of volume elements. In flowsheet simulation this simulation time is inexpedient. For the simulation with 400 elements, the relative error was $0.4\,\%$ and the calculation time varied in the range of 0.5 to 10 min, depending on the goodness of starting values. Therefore, a total number of elements $n_{\mathrm{total}} = 400$ is selected for all other simulations.

Table 7.1: Discretization accuracy of pyrolysis model

n_{total}	Y_{char}	Y_{oil}	Y_{gas}	E_{rel}
	wt.-%			%
100	19.50	51.54	30.57	1.6
200	19.47	51.11	30.22	0.8
300	19.46	50.97	30.11	0.5
400	19.45	50.90	30.05	0.4
2000	19.44	50.67	29.97	0.1

7.1.2 Superficial gas velocity (residence time)

The (initial) superficial gas velocity has an influence on the pyrolysis results. Primarily, the fluid dynamics, i.e. the dense bottom zone expansion and appearance, the suspension characteristics in the upper dilute zone and the entrainment rate at the reactor outlet depend upon the superficial gas velocity. Secondarily, the concentration profile, mass balances and residence times of gaseous and solid components and the product yields are affected. Therefore, the superficial gas velocity in the reactor itself is influenced, for instance when more gases evolve. Finally, the mass balance influences the fluid dynamics by the change of solid density and particle size and gas evolution. The effect of a lower mean solids density and larger particle size at lower superficial gas velocity (longer solids residence time) is included.

Figure 7.1 shows that for a fixed bed mass the dense bottom zone does expand higher at bigger u_0. The solids volume fraction in the dense bottom zone decreases and the slope of the exponential decrease in the dilute upper zone is smaller.

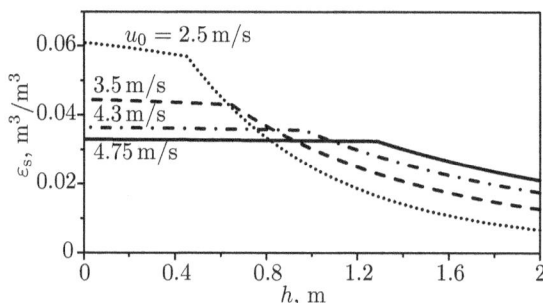

Figure 7.1: Solids volume fraction ε_s at different superficial gas velocities u_0

For the given reactor dimensions, the bubble diameter calculated by the used two-phase model [280–282] at a superficial gas velocity $u_0 > 4.3\,\mathrm{m/s}$ would be larger than and is thus limited to the reactor diameter. Therefore, in the dense bottom zone, a constant solids volume fraction is calculated for high u_0. With an increase in u_0, the higher drag force upon the particles induces not only the shallower exponential decay of the solids volume fraction in the dilute upper zone but also a higher solids entrainment rate. It is increased from about $10\,\mathrm{kg/(m^3 \cdot s)}$ at $2\,\mathrm{m/s}$ to about $60\,\mathrm{kg/(m^3 \cdot s)}$ at $5\,\mathrm{m/s}$.

The biggest influence of the superficial gas velocity is that the residence times of gas and solids in the reactor decrease substantially with the increase of u_0. The increase of superficial gas velocity from 2.5 to 4.5 m/s leads to a decrease in residence time of 63 % solids and 41 % gas (at 650 °C). This decrease results in a change in biomass (in this case lignin) conversion and product yield, depicted in Figure 7.2. A lower solid residence time (at higher u_0) makes a higher pyrolysis temperature necessary to obtain complete conversion. At higher superficial gas velocity the maximum oil yield is higher in value and shifted to higher temperatures, whereas the gas yield decreases. These trends can be explained by the lower vapor residence time leading to less secondary cracking of oil to gas. At a pyrolysis temperature below 525 °C, the char yield is higher at higher residence time as the conversion is more complete. Above a pyrolysis temperature of 525 °C, no influence of residence time can be observed as the lignin conversion is complete.

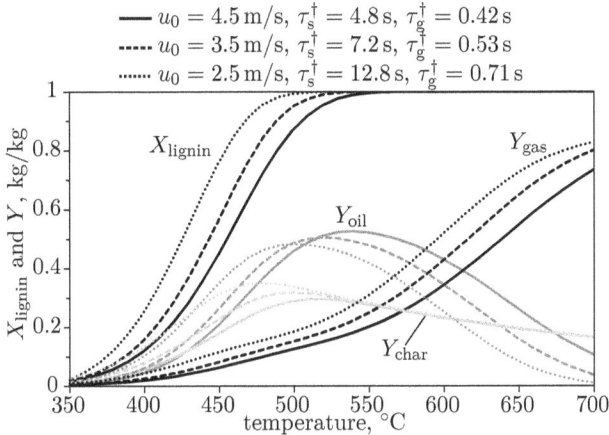

Figure 7.2: Lignin conversion X_{lignin} and product yields Y depending on superficial gas velocity u_0

† at 650 °C

7.1.3 Composition

The pyrolysis temperature influences the fluid dynamics and predominantly the reaction kinetics. With rising temperature the volume flow increases due to the heterogeneous solid-to-gas reactions and the secondary oil-to-gas reaction, as $M_{\text{oil}}/M_{\text{gas}} = 11.4$. The increase in superficial gas velocity due to the reaction is more distinct at higher temperature leading to a slightly higher bottom zone expansion. Also, the implemented change in bed material properties influences the fluid dynamics. The influence of temperature on biomass conversion and product yield is shown for the three pure components cellulose, hemicellulose, and lignin in Figure 7.3. As the chemical structure of the biomass components is fundamentally different also the pyrolysis reaction mechanism differs considerably. Therefore, also the product distribution of gas, oil, and char is essentially different at a certain temperature. The kinetic parameters (cf. Table 5.5) emulate this behavior.

As shown experimentally in a TGA (cf. Fig. 2.7) hemicellulose and lignin start to decompose at the lowest temperature of ~110 °C, whereas cellulose has a reaction onset of about 225 °C. Primary decomposition is complete at about 300 °C, 350 °C, and 450 °C for hemicellulose, cellulose, and lignin, respectively. Thus, lignin reacts over the broadest and cellulose over the narrowest temperature range. This behavior can also be observed

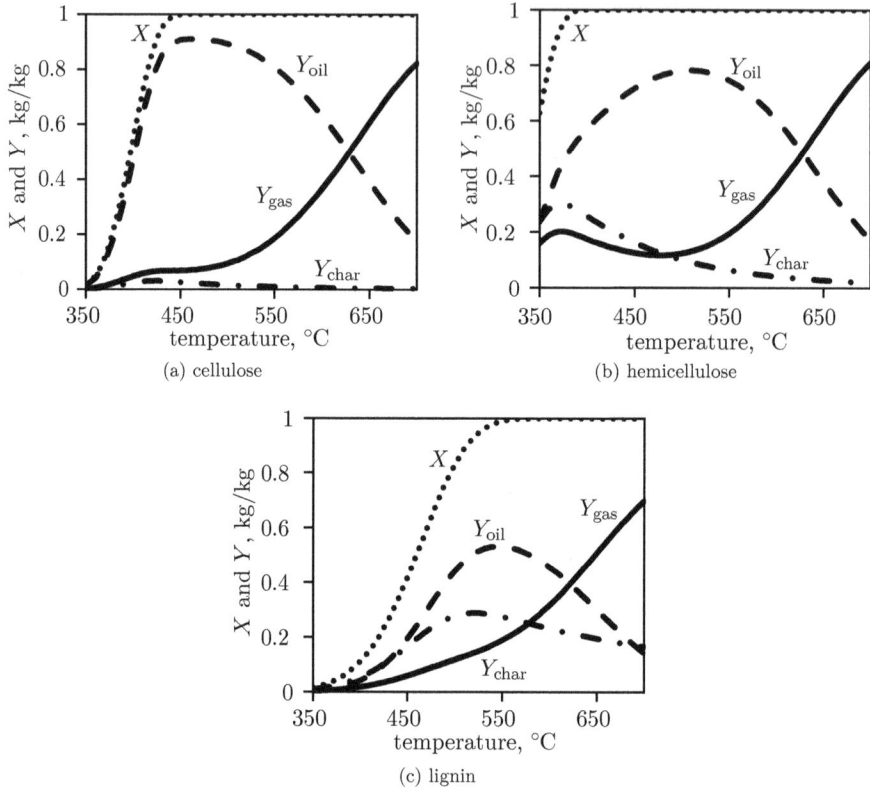

Figure 7.3: Biomass conversion X and product yields Y depending on pyrolysis temperature for main biomass components

for the modeled components in circulating fluidized bed pyrolysis, but at the higher heating rate and shorter residence time the temperature range is shifted to higher values of 360 to 390 °C, <350 to 440 °C and 360 to 560 °C for cellulose, hemicellulose, and lignin, respectively. It can also be observed that the decrease in oil yield from about 550 °C is steeper for cellulose than for hemicellulose and lignin. This model result is in good agreement with the experimental results of Qu et al. [102]. It can be explained by the nature of the pyrolysis oil of the three components: cellulose bio-oil mainly consists of levoglucosan, glycolaldehyde, ketones, and acids [102]. Levoglucosan yield can be up to 20 wt.-% at atmospheric pyrolysis conditions, 40 wt.-% for cellulose-particle vacuum pyrolysis and 58 wt.-% at vacuum pyrolysis conditions of cotton hydrocellulose [34]. Hemicellulose (xylan) pyrolysis oil contains mostly acids and furfural [102], whereas lignin pyrolysis oil contains mainly monomeric and oligomeric (polycyclic) aromatics, e.g. phenolic compounds. As levoglucosan has a lower temperature stability than furfural and phenols [349], the oil components are cleaved in the sequence cellulose, hemicellulose, and lignin. Additionally, the study of Liao et al. [350] supports this conclusion as a moderate pyrolysis temperature facilitates a high levoglucosan yield, whereas higher temperatures promote the formation of hydroxyacetaldehyde, formaldehyde, furfural, and acetol. The gas yield does in a general trend increase for all components as a higher temperature promotes the formation of gas via primary gas evolution and conversion of oil to gas via secondary reaction. The maximum gas yield is 83 wt.-% and 82 wt.-% for cellulose and hemicellulose at 700 °C. Lignin has the lowest gas yield, which e.g. is 70 wt.-% at the same temperature. Below 550 °C the gas yield of hemicellulose is higher than that of the other two biomass components, but above it is outperformed by the cellulose gas yield. This simulation result is in good agreement with the experimental findings of Qu et al. [102], who found the interception of cellulose and hemicellulose gas yield to occur at 560 °C. Char formation shows for all three components a maximum at low pyrolysis temperatures from where the yield gradually decreases as the carbonization is favored at higher temperature. For cellulose and hemicellulose, a very low amount of char in a single-digit magnitude is formed. For cellulose, this result is in good agreement with the literature: the final char yield of cellulose rapid pyrolysis at high temperature is 6 wt.-% [351]. Lignin has a much higher char yield than the other two components of 17 wt.-% at 700 °C (cf. also TGA measurement in Figure 2.7).

7.2 Granulation

For Kraft lignin, it was shown (cf. Section 6.1.1) that the bed material is coated with char during pyrolysis. As char and sand have a different density and the particles grow through the increase in char layer, the fluid dynamics are influenced. The model depends on the mean particle diameter and the density of the char-sand-mixture. The char layer thickness mainly depends on the solids residence time at fixed biomass feeding mass flow. According to Eq. 5.45 the solids residence time depends on the solids entrainment rate. The entrainment rate correlation [293] returns a value of 17.35 kg/(m$^2 \cdot$s) for the SPE pilot plant at the given parameters (Table A.1). For the riser solids holdup of 0.7 kg, the model returns a solid residence time of $\tau_s = 5.4$ s and a char layer thickness of 0.07 μm at 650 °C. To validate this modeling result, it is necessary to convert the calculated char layer thickness to experimental conditions. In the experiments the char amount accumulates over the experiment duration on the initial bed material, which is circulating many times through the lignin spray zone. For instance in experiment V81, the initial bed material of 5 kg quartz sand with a sauter diameter of 175 μm remained for 130 min in the pyrolysis process. In contrast, quartz sand is fed continuously into the reactor in the stationary model. The bed material is thus only passing the riser in one cycle (with according residence time distribution) before exiting the reactor. A proportional relation is true for char mass and the residence time:

$$\frac{m_{char,model}}{m_{char,exp}} = \frac{\tau_{s,model}}{\tau_{s,exp}} \quad . \tag{7.2}$$

The char mass of both experiment and model can further be defined by the product of total particle number $n_p = m_{QS}/(\pi/6 \cdot d^3_{sauter, QS} \cdot \rho_{a, QS}) = 3.6 \cdot 10^8$ with the mass of a char particle layer:

$$m_{char} = n_p \cdot \frac{\pi}{6} \cdot \rho_{char} \cdot \left(\bar{d}^3_{sauter} - d^3_{sauter,QS} \right) \quad . \tag{7.3}$$

In the experiment, the holdup of the total CFB system has to be considered, whereas in the model only the riser holdup is accounted for. Therefore, the Eq. 7.4 can be substituted for $m_{char,model}/m_{char,exp}$.

$$\frac{m_{char,model}}{m_{char,exp}} = \frac{m_{QS,model}}{m_{QS,exp}} \cdot \frac{d^3_{sauter,QS,exp}}{d^3_{sauter,QS,model}} \cdot \frac{\left(\bar{d}^3_{sauter,model} - d^3_{sauter,QS,model} \right)}{\left(\bar{d}^3_{sauter,exp} - d^3_{sauter,QS,exp} \right)} \tag{7.4}$$

By equating the right-hand sides of equations 7.2 and 7.4 an equation for comparison of experimental and model char layer thickness is obtained. It holds

$$\frac{\left(\bar{d}^3_{\text{sauter,model}} - d^3_{\text{sauter,QS,model}}\right)}{\left(\bar{d}^3_{\text{sauter,exp}} - d^3_{\text{sauter,QS,exp}}\right)} = \frac{\tau_{\text{s,model}}}{\tau_{\text{s,exp}}} \cdot \frac{m_{\text{QS,exp}}}{m_{\text{QS,model}}} \cdot \frac{d^3_{\text{sauter,QS,model}}}{d^3_{\text{sauter,QS,exp}}} \quad . \tag{7.5}$$

Using Eq. 7.5 the char layer of $0.7\,\mu\text{m}$ in the model can be converted to $10.6\,\mu\text{m}$ at experimental conditions. This value is in good agreement with the measured values for experiment V81. The bed material of experiment V81 having a mean char layer thickness of $6.5\,\mu\text{m}$ lying in between 4 to $13\,\mu\text{m}$ (cf. Figure 6.11).

7.3 Yields & conversion

7.3.1 Validation of applicability

For model validation, experiments with different biomass types and lignin are simulated for three different biomass pyrolysis plants: CFB pyrolysis of pine wood at CERTH, BFB pyrolysis of wood and straw at the University of Waterloo and CFB Kraft and hydrolysis lignin pyrolysis obtained in this work. The model as described in Chapter 5 uses, in particular, the kinetics of Miller and Bellan [221]. The comparison between model and experiment is assessed by relative error defined as

$$E_{\text{rel}} = \frac{\left[Y^{\text{model}} - Y^{\text{exp}}\right]}{Y^{\text{exp}}} \cdot 100\,\% \quad . \tag{7.6}$$

CFB pyrolysis of lignin

The modeled yield curves for pyrolysis of Kraft and hydrolysis lignin are depicted in Figures 7.4a and 7.4b, respectively. For both materials, the experimental oil yield is scaled so that the sum of gas, char and oil yield is $100\,\text{wt.-\%}$. This approach can be explained by the fact that the incomplete recovery in the experiments is mainly water (cf. Section 6.2.1.2 "Recovery rate and error discussion"), which is the main constituent of the watery oil fraction and thus part of the model oil yield. Furthermore, for hydrolysis lignin, the measured char yield is corrected with the ash content in the hydrolysis lignin char. It is with about $50\,\text{wt.-\%}$ higher than in Kraft lignin char ($6\,\text{wt.-\%}$) and results from the considerably higher ash content in the hydrolysis lignin ($11.3\,\text{wt.-\%}$, obtained from straw with ash content of $6.4\pm2.2\,\text{wt.-\%}$ [14, 28, 29, 100, 133, 189, 352–355]) compared to Kraft lignin ($1.1\,\text{wt.-\%}$) from woody biomass (ash content of 0.19 to $0.85\,\text{wt.-\%}$ [14, 51, 58, 169, 189, 231, 352, 356, 357]). It can be seen that for both lignins the simulated char yield development is in agreement with the experimental values. However, the model results for lignin pyrolysis do not agree well with the experimental yields of pyrolysis oil and gas. The pyrolysis oil yield is predicted too low for both lignins. The deviation increases with rising pyrolysis temperature. Inversely, the gas yield is increasingly overestimated by the model with the rise in temperature.

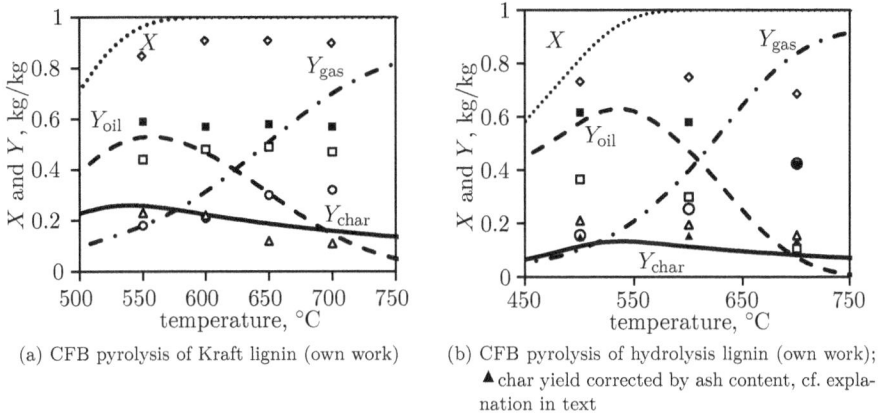

(a) CFB pyrolysis of Kraft lignin (own work)

(b) CFB pyrolysis of hydrolysis lignin (own work); ▲ char yield corrected by ash content, cf. explanation in text

Figure 7.4: Biomass conversion X and product yields Y: Comparison of model with experimental results; model: lines with literature kinetics ($k_{4,0,\text{lit}}^{\text{lignin}} = 4.28 \cdot 10^6$/s) [221] and experimental: ◇ conversion, □ oil yield, ■ oil yield scaled to 100 wt.-% conversion, ○ gas yield, and △ char yield

CFB pyrolysis of pine wood

The yields and conversion of pine wood pyrolysis in the autothermal pyrolysis system at CERTH are compared to the model. The model input parameters are adjusted to match the experimental setup of the autothermal pyrolysis system at CERTH, which is described in Section 5.3.1 and the simulation parameters, which are listed in Table A.2. For comparison of the simulation results with the reported experiments, especially the experiments R10, R9, and R8 are relevant because of the biomass feed particle size of 1 to 1.5 mm. The other two experiments (R11 and R7) were carried out with particles in the size range 1.5 to 2 mm, for which intra-particle diffusion and temperature gradient play a greater role. The comparison for the pine wood chip particles with a size of 1 to 1.5 mm is shown in Figure 7.5. In general, the model describes the yield development with temperature as determined in the experiment. The accuracy is better for lower temperature. The char yield shows the correct trend and magnitude but is predicted too low by the model for all experiments.

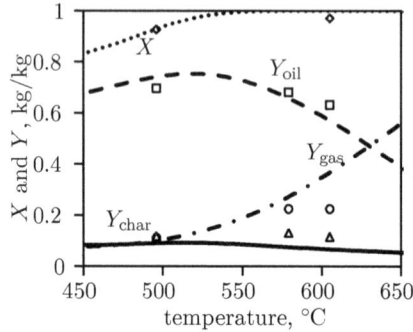

Figure 7.5: Biomass conversion X and product yields Y for CFB pyrolysis of pine wood [89]: Comparison of model with experimental results; model: lines with literature kinetics ($k_{4,0,\text{lit}}^{\text{lignin}} = 4.28 \cdot 10^6$/s) [221] and experimental: $\diamond$ conversion, $\square$ oil yield, $\blacksquare$ oil yield scaled to 100 wt.-% conversion, $\bigcirc$ gas yield, and $\triangle$ char yield

BFB pyrolysis of wood and straw

For further model validation experimental results of Scott and Piskorz [84] are used. The experiments were carried out in a bubbling fluidized bed with a superficial gas velocity ranging from 0.54 to 0.81 m/s. Wheat straw and two wood types, i.e. maple and standard poplar-aspen were pyrolyzed. Thus, it is possible to investigate the model performance for a broad biomass composition and a different fluidized bed regime. The experimental setup, including plant dimensions and operating parameter, at the University of Waterloo is described in detail in Section 5.3.2. Exemplary the results for maple wood are shown in Figure 7.6, the other two biomass types (standard poplar-aspen and wheat straw) are discussed in detail in Section 7.3.2. Maple wood has a quite different biomass composition compared to pine wood (cf. Tables 5.7 and 5.9) and also the fluidized bed is operated at a superficial gas velocity with a considerably smaller magnitude compared to the autothermal CFB riser at CERTH ($u_0 = 3.85$ to 5.31 m/s). Still, the yields of all pseudo-products are predicted in the range of 474 to 530 °C with a visibly small error.

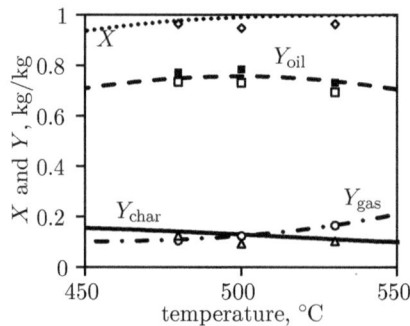

Figure 7.6: Biomass conversion X and product yields Y for BFB pyrolysis of maple wood [84]: Comparison of model with experimental results; model: lines with literature kinetics ($k_{4,0,\text{lit}}^{\text{lignin}} = 4.28 \cdot 10^6$/s) [221] and experimental: $\diamond$ conversion, $\square$ oil yield, $\blacksquare$ oil yield scaled to 100 wt.-% conversion, $\bigcirc$ gas yield, and $\triangle$ char yield

Conclusion

In summary, it can be seen that for woody biomass with a lignin content of 23 wt.-% (pine wood) and 22 wt.-% (maple wood) the model does agree much better with the experimental results than for lignin with a purity of 95.43 wt.-% and 48.85 wt.-% (Kraft and hydrolysis lignin, respectively). For high lignin content, especially the gas yield is over- and oil yield accordingly underestimated by the model. The literature model kinetics [221] assume the same kinetics for the secondary reaction, i.e. conversion of oil to gas, for all biomass components. This assumption is questionable, as the pyrolysis oil composition is fundamentally different for lignin-derived oil compared to oil from carbohydrates [102]. Furthermore, it was shown that the pyrolysis oils of cellulose and hemicellulose have a lower temperature stability than lignin-derived pyrolysis oil [349]. It can thus be deduced that the kinetics of the secondary reaction are not suitable for simulation of pyrolysis of biomass with high lignin content.

7.3.2 Kinetic improvement for lignin

The model improvement approach for the above-explained deficiency of the lignin-derived oil and gas yields is to fit the kinetic constant $k_{4,0}^{\text{lignin}}$ of the secondary reaction, converting oil to gas, to the experimental data of Kraft lignin pyrolysis. The Kraft lignin pyrolysis experiments are chosen because of the high purity of about 95.5 wt.-%. The objective error function E for finding $k_{4,0}^{\text{lignin}}$ is defined as the minimum of the sum of quadratic errors between experimental and simulated gas and oil yields.

$$
E(k_{4,0}^{\text{lignin}}) = \min \left(\left[\sum_{i}^{n} (Y^{\text{exp}} - Y^{\text{model}})^2 \right]_{\text{gas}} + \left[\sum_{i}^{n} (Y^{\text{exp}} - Y^{\text{model}})^2 \right]_{\text{oil}} \right) \tag{7.7}
$$

Only the kinetic constant for the secondary reaction of lignin is adjusted, while the kinetics for cellulose and hemicellulose remain unchanged. With the new obtained kinetic constant the simulations are repeated for all compared experimental data.

CFB pyrolysis of lignin

Fitting the data of Kraft lignin to the scaled oil yield (oil and water) gives a kinetic constant of $k_{4,0,\text{fit}}^{\text{lignin}} = 1.15 \cdot 10^6/\text{s}$. The found solution is compared to the literature kinetics ($k_{4,0,\text{lit}}^{\text{lignin}} = 4.28 \cdot 10^6/\text{s}$) and the experimental data in Figure 7.7a. Additionally, the relative errors for both kinetic data sets are listed in Table 7.2. The char yield does not change as only the secondary reaction kinetics are adjusted. Through deceleration of the secondary reaction the sum of relative errors for the gas yield is reduced from 60 below 25 %. Furthermore, the relative error of the oil yield is decreased from 38 to less than 7.5 %. The new fit shifts the oil yield maximum from 560 °C to a higher temperature of about 600 °C.

For hydrolysis lignin (cf. Figure 7.7b) a good agreement is achieved for pyrolysis oil yield with the obtained kinetics. The relative error is reduced by up to 400 %. The gas yield is underestimated at low pyrolysis temperatures and overestimated at high temperature. Compared to the literature kinetics a much better agreement between model and experimental data is achieved with the new kinetic constant $k_{4,0,\text{fit}}$. The oil yield maximum is also shifted to a higher temperature, i.e. by about 25 °C to 560 °C. The temperature of the maximum oil yield is lower than for Kraft lignin due to the higher content of cellulose and hemicellulose, which degrade at a lower temperature. On the other hand, the maximum is about 3 wt.-% higher for hydrolysis lignin than for Kraft lignin due to higher oil yields obtained from pyrolysis of carbohydrates compared to lignin.

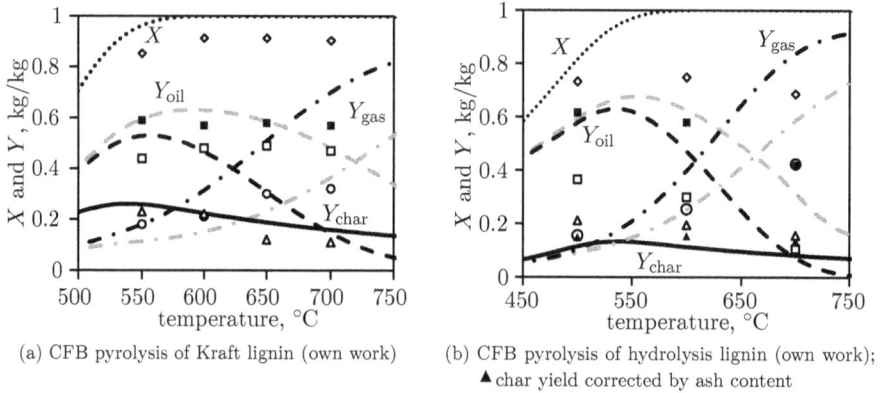

(a) CFB pyrolysis of Kraft lignin (own work)

(b) CFB pyrolysis of hydrolysis lignin (own work); ▲ char yield corrected by ash content

Figure 7.7: Biomass conversion X and product yields Y: Comparison of model with experimental results; model: black lines literature kinetics ($k_{4,0,\text{lit}}^{\text{lignin}} = 4.28 \cdot 10^6$/s) [221], gray lines fitted kinetic value of $k_{4,0,\text{fit}}^{\text{lignin}} = 1.15 \cdot 10^6$/s and experimental: ◇ conversion, □ oil yield, ■ oil yield scaled to 100 wt.-% conversion, ○ gas yield, and △ char yield

Table 7.2: Lignin pyrolysis: relative errors between simulation and experiment with literature kinetics ($k_{4,0,\text{lit}}^{\text{lignin}} = 4.28 \cdot 10^6$/s) [221] and fitted kinetic value of $k_{4,0,\text{fit}}^{\text{lignin}} = 1.15 \cdot 10^6$/s

lignin type	ϑ	E_{rel}, %				
		gas		oil		char
	°C	lit	fit	lit	fit	lit = fit
Kraft lignin	550	4.1	-34.5	-9.9	3.5	38.8
Kraft lignin	600	49.9	-27.6	-19.9	10.2	30.9
Kraft lignin	650	68.8	-22.8	-47.9	0.8	19.3
Kraft lignin	700	117.0	13.9	-74.7	-15.7	2.2
hydrolysis lignin	500	-34.3	-42.5	-6.2	-3.5	-8.1
hydrolysis lignin	600	56.5	-1.2	-18.7	7.8	-12.2
hydrolysis lignin	700	97.4	38.5	-82.8	-20.8	-29.8

CFB pyrolysis of pine wood

The simulation for pyrolysis of pine wood in the CFB system of CERTH was repeated with the fitted kinetics. The resulting yield errors are compared to literature kinetics in Table 7.3. Furthermore, the results are compared to the three experiments with 1 to 1.5 mm pine wood chip particle size in Figure 7.8. Both gas and oil yields are represented in better way by the new obtained kinetic parameter set. Simulating pine wood chip particles with larger diameter results in a bigger error for the oil yield, which is overestimated by the model. The larger deviation can be explained by the inter-particle effects, taking place at a larger feed particle size, which are not considered in the model. Especially, the particle (core) temperature is lower for larger particles (larger Bi number). Thus, a less rapid devolatilization and enhanced cracking of oil to gas occur [131, 134].

Table 7.3: CFB pyrolysis [89] of pine wood: relative errors between simulation and experiment with literature kinetics ($k_{4,0,\text{lit}}^{\text{lignin}} = 4.28 \cdot 10^6/\text{s}$) [221] and fitted kinetic value of $k_{4,0,\text{fit}}^{\text{lignin}} = 1.15 \cdot 10^6/\text{s}$

| Exp. | d_p | ϑ | E_{rel}, % | | | | char |
| | | | gas | | oil | | |
	mm	°C	lit	fit	lit	fit	lit = fit
R10	1-1.5	496	-11.8	-13.2	6.0	6.5	-23.3
R9	1-1.5	579	22.4	11.0	-3.8	-0.2	-43.0
R8	1-1.5	605	65.9	47.5	-10.3	-3.9	-42.0
R11	1.5-2	550	-5.1	-11.4	23.8	26.4	-27.2
R7	1.5-2	581	-28.3	-35.0	34.2	39.4	-42.1

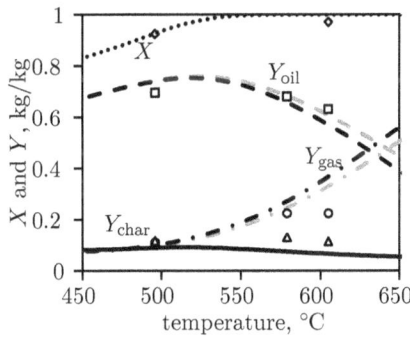

Figure 7.8: Biomass conversion X and product yields Y for CFB pyrolysis of pine wood [89]: Comparison of model with experimental results; model: black lines literature kinetics ($k_{4,0,\text{lit}}^{\text{lignin}} = 4.28 \cdot 10^6/\text{s}$) [221], gray lines fitted kinetic value of $k_{4,0,\text{fit}}^{\text{lignin}} = 1.15 \cdot 10^6/\text{s}$ and experimental: $\diamond$ conversion, $\square$ oil yield, $\blacksquare$ oil yield scaled to 100 wt.-% conversion, $\bigcirc$ gas yield, and $\triangle$ char yield

BFB pyrolysis of wood and straw

For maple wood (Fig. 7.9a) with a composition of 40 wt.-% cellulose, 38 wt.-% hemicellulose and 22 wt.-% lignin the yield of gas and oil and therefore also the relative deviation between experiment and model do not change decisively in the considered temperature region (cf. Table 7.4). With the kinetic constant of the secondary reaction for the conversion of lignin oil to gas, the same good agreement is achieved.

(a) BFB pyrolysis of maple wood [84]

(b) BFB pyrolysis of standard poplar-aspen [84]; ● gas yield at 625 °C scaled to 100 wt.-% conversion

(c) BFB pyrolysis of wheat straw [84]; ▲ char yield corrected by ash content

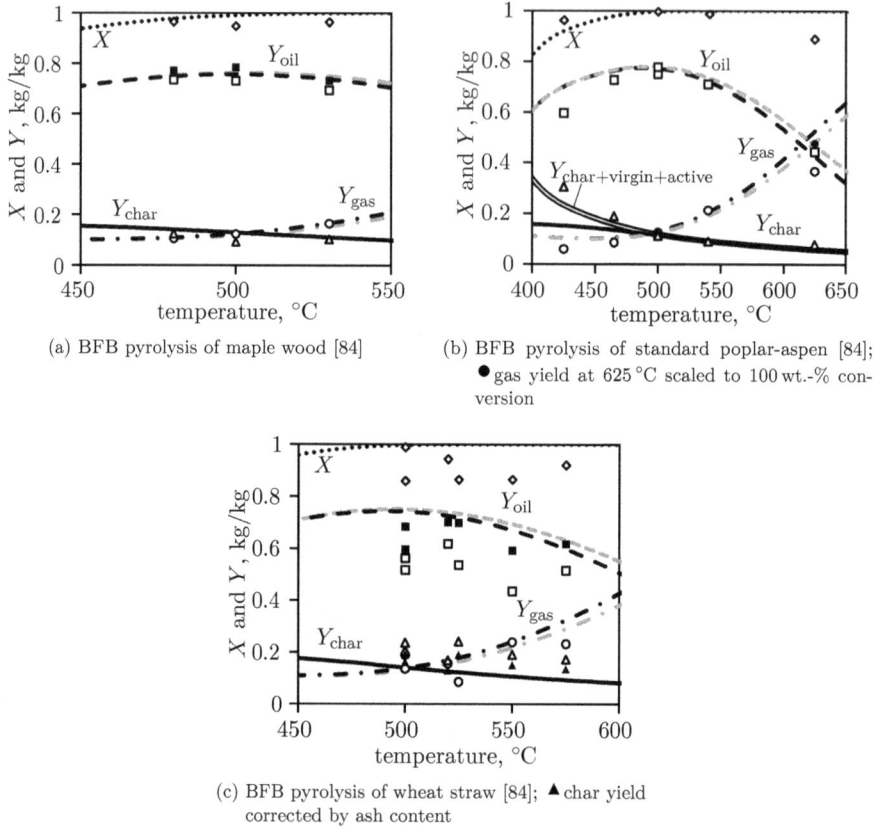

Figure 7.9: Biomass conversion X and product yields Y: Comparison of model with experimental results; model: black lines literature kinetics ($k_{4,0,\text{lit}}^{\text{lignin}} = 4.28 \cdot 10^6 /\text{s}$) [221], gray lines fitted kinetic value of $k_{4,0,\text{fit}}^{\text{lignin}} = 1.15 \cdot 10^6 /\text{s}$ and experimental: ◇ conversion, □ oil yield, ■ oil yield scaled to 100 wt.-% conversion, ○ gas yield, and △ char yield

Figure 7.9b shows the comparison of model and experiments for standard poplar-aspen with a composition of 50.2 wt.-% cellulose, 31.6 wt.-% hemicellulose and 18.2 wt.-% lignin. The experimental char yield is represented very well for $\vartheta \geq 500\,°\text{C}$. At lower pyrolysis temperature the experimental char yield is higher than the predicted values. This deviation can be explained by incomplete conversion of aspen-wood particles at these temperatures, resulting in a solid residue that still contains considerable amounts of biomass that is not completely converted. However, the solid residue of pyrolysis experiments below $\vartheta \geq 500\,°\text{C}$ is in good agreement with the curve of $Y_{\text{char+virgin+active}}$ depicted also in Figure 7.9b. The gas yield is overestimated by the model at temperature below 470 °C. At 625 °C the experimental gas yield is lower than both model predictions. Comparing also the oil yield and the overall conversion X at 625 °C, it can be observed that for oil there is good agreement between model and experiment and the recovery rate is about 10 wt.-% lower than complete conversion as should be expected at this temperature. Thus, if the gas yield was wrongly determined in the experiment to be 10 wt.-% too low also good agreement for the gas yield would be achieved at 625 °C (cf. ● in Figure 7.9b). In the medium temperature range of 450 to 550 °C the measured oil yield is in good accordance with the experimental data. Overall it can be concluded that at $\vartheta \geq 450$ °C there is good agreement for both kinetics data sets.

Table 7.4: BFB pyrolysis [84]: relative errors between simulation and experiment with literature kinetics ($k_{4,0,\text{lit}}^{\text{lignin}} = 4.28 \cdot 10^6$/s) [221] and fitted kinetic value of $k_{4,0,\text{fit}}^{\text{lignin}} = 1.15 \cdot 10^6$/s

biomass type	ϑ	E_{rel}, %				
		gas		oil		char
	°C	lit	fit	lit	fit	lit = fit
maple wood	480	4.7	0.5	-2.9	-2.4	14.2
maple wood	500	1.0	-4.1	-3.6	-2.8	39.9
maple wood	530	0.7	-6.6	0.8	2.5	7.5
wheat straw	500	-23.1	-31.2	22.4	25.7	-29.6
wheat straw	500	-2.1	-4.4	8.5	9.7	-19.2
wheat straw	520	0.3	-0.5	4.2	5.4	-7.9
wheat straw	525	86.9	90.6	4.5	5.0	-42.8
wheat straw	550	7.2	-8.9	9.0	17.1	-31.2
wheat straw	575	25.2	25.2	0.3	2.2	-34.2
aspen wood	425	76.0	71.8	10.8	10.1	-48.5
aspen wood	465	24.4	19.7	5.3	5.5	-27.4
aspen wood	500	2.5	-0.5	2.6	3.2	-6.7
aspen wood	500	20.2	2.6	14.6	1.4	15.0
aspen wood	540	-4.0	-11.0	3.3	5.4	1.5
aspen wood	625	46.6	31.4	-25.3	-15.3	-24.9

The results for wheat straw (composition: 38.2 wt.-% cellulose, 48.4 wt.-% hemicellulose, and 23.4 wt.-% lignin) are compared in Figure 7.9c. It can be seen that the measurement uncertainty is higher for wheat straw than for woody biomass. Also, the deviation between experimental data and model prediction is thus higher and therefore a greater uncertainty persists for applicability of the model. For comparison – analogously to lignin pyrolysis – the measured oil yield is scaled to achieve full conversion. Again, this approach can be justified by the higher probability of inaccuracy of oil yield measurement. With this assumption the oil yield has the same magnitude as predicted by the models (with both kinetics). The relative error given in Table 7.4 is slightly higher than for woody biomass. The char yield is predicted too low because of the high ash content of the wheat straw (6.4±2.2 wt.-% [14, 28, 29, 100, 133, 189, 352–355]) compared to e.g. pine wood 0.19 to 0.85 wt.-% [58, 169, 356, 357], aspen wood 0.39 wt.-% [51] or maple wood 0.59 wt.-% [231]. This results in an ash content in the char of 19.07 to 22.65 wt.-%. If corrected to a char content of woody biomass the agreement between model and experiment is increased. Nevertheless, it was shown that the minerals content, which is not considered by the applied biomass kinetics (additivity law for biomass components), should be considered [225]. As the minerals content for the examined biomass is highest in straw, the predicted yield error are bigger than for the other biomasses.

Conclusion

The conversion by the secondary reaction from lignin-oil to gas was overestimated in the literature. Therefore, the pre-exponential factor of the kinetic constant was adjusted to accurately describe Kraft lignin pyrolysis. Hence, this adjustment resulted in significant reduction of relative error for both gas and oil yield. Furthermore, an improved agreement between model and experimental data is achieved for hydrolysis lignin, while the conformity is sustained for pyrolysis of lignocellulosic biomass (wheat straw, aspen, maple, and pine wood). The accuracy is lower for straw and straw derived hydrolysis lignin than for the other investigated biomasses. The lower accuracy might be explained by the different chemical structure of biomass components in straw compared to biomass. Exemplary straw lignin consists of GSH functional groups, whereas soft and hardwood consist of G and GS groups, respectively [29–31]. Furthermore, as char and mineral matter influence the pyrolysis behavior and the resulting product yields [161, 165–167] further research is necessary to integrate this effect especially for biomasses with high ash content, like straws.

7.4 Energetic evaluation of an integrated pyrolysis-combustion process

Flowsheet simulation of the overall pyrolysis process, including by-product combustion (cf. Fig. 5.8), leads to valuable information on the energetic feasibility of the process. For evaluation the energetic performance ratio η_{SD}, it is compared to the performance of the pyrolysis process itself. This pyrolysis process performance can be expressed by the oil energy recovery rate η_{OR}, i.e. the feed specific energy content in the main product (oil). Energy sufficiency is achieved if the combustion of gas and char yields enough energy for the auxiliary heat consumption of the pyrolysis plant. Favorably would be a process which achieves a high energy content in the oil and is self-sufficient in terms of process energy. However, these goals are contradictory, but an optimum can be aspired. In the following, the influence of the process parameters: superficial gas velocity in the pyrolysis riser u_0, lignin feeding rate $\dot{m}_L$ and pyrolysis temperature ϑ_{PR} is discussed.

Figure 7.10a (a) depicts the dependency of the two performance ratios on the superficial gas velocity u_0, which was varied in the range of 2.5 to 5 m/s. The dotted line at $\eta_{SD} = 1$ represents the limit below which supplemental energy has to be introduced into the process, e.g. by gas burners. Above $\eta_{SD} = 1$ the pyrolysis process is self-sufficient. With increasing superficial gas velocity the energy recovery rate η_{OR} increases significantly, whereas the overall energetic process performance ratio η_{SD} decreases notably. These trends can be mainly explained by the decreasing pyrolysis gas and vapor residence time in the riser. Thus, secondary oil cracking reactions are suppressed, leading to a higher oil yield, which is proportional to η_{OR}. The increase in oil yield is equivalent to a decrease in gas and char. Hence η_{SD} decreases to fall below the limit of $\eta_{SD} = 1$ at $u_0 = 3.67$ m/s.

Figure 7.10 (b) provides the dependency of the two performance ratios on pyrolysis temperature ϑ and lignin feed $\dot{m}_L$. The mean superficial gas velocity over the riser height is kept constant at $u_0 = 4$ m/s, i.e. the fluidizing gas supply is decreased with increasing evolution of gaseous products caused by an increasing lignin mass flow $\dot{m}_L$. Therefore, the influence on η_{OR} – especially at pyrolysis temperature below 600 °C – is negligible. The surplus-deficit-ratio η_{SD} has a minimum for all lignin feeding rates. At temperatures below 500 °C the ratio increases with decreasing pyrolysis temperature due to an incomplete conversion. Thus, instead of conversion to pyrolysis products, an increasing amount of virgin and active biomass is combusted. At temperatures higher than $650\,°C$ η_{SD} increases due to an increasing permanent gas yield. The minimum of η_{SD} concurs with the maximum of η_{OR}, which is about 0.5 at $600\,°C$. Thus, for achieving both a large energy amount in the liquid product and process operation in the autothermal regime the lignin feeding rate has to be increased. In this case, autothermal operation is achieved at feeding rates of $15\,kg/h$ or higher. Though this mass flow is larger than the ones investigated experimentally in this work, studies exist with higher biomass throughput in the circulating fluidized bed riser. Freel and Graham [108] successfully pyrolyzed biomass at throughput of 0.46 to $2.2\,kg/(m^2 \cdot s)$. The corresponding value of $0.83\,kg/(m^2 \cdot s)$ for a feeding rate of $15\,kg/h$ falls within this range. Therefore, a successful autothermal operation at lignin feeding rates greater than $15\,kg/h$ is likely. But due to increased agglomeration tendency at higher throughput, particular care on the selection of the heat-carrier-to-biomass feeding ratio should be taken.

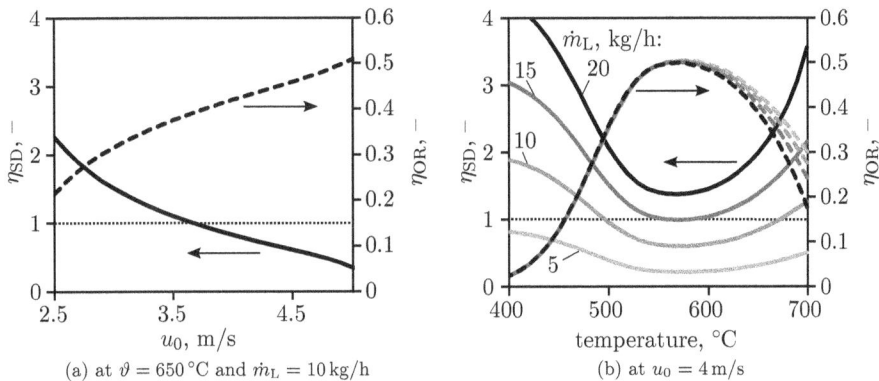

(a) at $\vartheta = 650\,°C$ and $\dot{m}_L = 10\,kg/h$

(b) at $u_0 = 4\,m/s$

Figure 7.10: Energetic performance ratios for Kraft lignin pyrolysis with integrated by-product combustion

8 Summary and conclusions

In search of an alternative for energy and chemicals from fossil fuels, for the first time, an in-depth study on lignin pyrolysis in a circulating fluidized bed reactor was carried out. This process promises high liquid yields due to short vapor residence time in the hot reaction zone. A pyrolysis plant consisting of pneumatic feeding system, circulating fluidized bed reactor, secondary cyclone, scrubber, and after burning chamber was designed, erected and operated. For examination of deviations in pyrolysis behavior, a softwood Kraft lignin and a wheat straw hydrolysis lignin have been pyrolyzed.

Morphological analysis of the bed material from CFB lignin pyrolysis, together with thermal characterization of the lignins and biomass reference materials in a Flash-DSC analyzer, were conducted. This novel investigation gave insight into the fundamental differences in pyrolysis behavior. It could be deduced, that a higher impurity with cellulose and hemicellulose inhibits lignin melting. Therefore, hydrolysis lignin does keep its overall particle shape during pyrolysis, i.e. it reacts in form of a solid matrix and devolatilization occurs from this matrix. The char, which might be used as molecular sieve, is rich in micro, meso, and macro pores. For application as activated carbon, further enlargement of the chars' specific surface area by steam or CO_2 is necessary. On the contrary, Kraft lignin melts and spreads on the bed material particles to react on the surface. Devolatilization occurs while the intermediate layer is molten. Thus, if a bubble of volatiles is released from the liquid layer the surface tension evens out the surface, resulting in a virtually macro pore-free char layer with comparatively low porosity. It could be found that the growth rate of a char layered bed material particle depends mainly on the lignin feed rate and the instantaneous char yield, which itself is depending linearly on the amount of char in the reactor. Furthermore, also the particle size of the bed material does have a linear influence on the granulation, which accelerates with increase in particle size. It should be remarked that understanding this different behavior is crucial for further modeling and especially for catalytic pyrolysis processes, as the Kraft lignin char will deactivate a solid catalyst in shorter time.

The pyrolysis performance was determined by measurement of the overall yields of gas, oil, and char as well as their composition. The optimal oil yield for Kraft lignin was determined in the temperature range of 600 to 650 °C with about 50 wt.-% excluding and 60 wt.-% including reaction water. The pyrolysis oil yield of hydrolysis lignin has – in the analyzed temperature range at 500 °C – a maximum of 36.5 wt.-% without and 60 wt.-% with reaction water. It could be demonstrated that the obtained oil yields are higher than in other pyrolysis equipment such as fixed bed or centrifuge reactors and most reported fluidized beds. The organic (non-water) oil consists of approximately 20% monomeric substances of which more than 340 have been identified and quantified. Kraft lignin oil contains mainly aromatic structures, whereas hydrolysis lignin oil includes also various non-aromatic substances, which to large degree originate from the carbohydrate impurities. Both lignin oils have a high O/C-ratio compared to crude oil. This high oxygen content is the major hurdle for technical application. Therefore, further research on catalytic upgrading techniques such as stabilization by (hydro)deoxygenation, hydrogenation, and condensation reactions of oxygenated compounds is necessary. Due to the higher carbohydrate content, the gas yield of hydrolysis lignin is 3 to 10 wt.-% larger than for Kraft lignin. The maximum is 42.4 wt.-% and 32.7 wt.-% at 700 °C for hydrolysis and Kraft lignin, respectively. The char yield has a magnitude of about 20 wt.-% for both lignins, decreasing with increasing temperature.

The mineral matter contained in the lignin accumulates in the char holdup of the CFB. The inorganics are catalytically active and influence product yields and composition. As a result char and gas yields increase, whereas the oil yield decreases. The molecular weight distribution of the oil oligomers is due to this influence lightly shifted to smaller values. From the development of monomer yields, it can be deduced that the inorganics favor dehydration reactions and $C-O$ scission, promote demethoxylation, decarboxylation, and scission of $C-C$ bonds. The degree of demethylation is diminished. These reactions promote also an overall increase of CO and CO_2 evolution. The experimental work can finally be concluded by the fact that – despite the agglomeration, defluidization, and clogging issues reported in literature – pyrolysis of sticky lignin can be executed in a circulating fluidized bed resulting in high liquid yield. A catalyst-screening for a catalyst with high a high monomer-selectivity is desirable. Furthermore, the performance of a reactor-regenerator-looping process with such a catalyst should be determined.

The experimental work was accompanied by simulation of the biomass pyrolysis process. Therefore, a new semi-empirical pyrolysis model for the fluidized bed reactor was developed and implemented as a flowsheet unit model in the Aspen Custom Modeler ACM®. The model describes the fluid dynamics of the fluidized bed in a dense bottom and a dilute upper zone as well as the micro-particle pyrolysis reactions and mass balances. Model validation has been carried out with diverse biomasses in different fluidized bed systems (bubbling and circulating fluidized bed). The literature model kinetics questionable assume the same value for the secondary reaction converting oil to gas for all biomass components. But with the implemented literature kinetics the lignin gas yield is over- and oil yield accordingly underestimated. Therefore, the secondary reaction kinetics for lignin was refitted with the experimentally obtained Kraft lignin data. The adjustment resulted in significant accuracy improvement for both gas and oil yield and for every modeled biomass composition (especially the lignins). Thus, a model which is predicting pyrolysis process yields for a broad pyrolysis feedstock composition (by means of cellulose, hemicellulose and lignin content) and is applicable to both circulating and bubbling fluidized beds was deduced.

For energetic evaluation of the pyrolysis process, the derived model was implemented in a flowsheet plant model with integrated char and permanent gas combustion. The plant model returns the mass and heat flow between all connected flowsheet unit operations. Based on the results the energetic performance of the system (described by the surplus-deficit-ratio) and the pyrolysis performance (described by the liquid product energy recovery rate) were compared. The simulation shows that the minimal surplus-deficit-ratio concurs with the maximal liquid product energy recovery rate, which is about 0.5 at 600 °C. To achieve operation in the autothermal regime, while maintaining that maximum liquid product energy recovery rate, at pilot scale a lignin feeding rate greater than 15 kg/h is necessary.

To summarize, this comprehensive work on lignin pyrolysis covers thorough investigations on pyrolysis mechanism, product yields in a CFB system as well as process modeling. It broadens knowledge on CFB lignin pyrolysis and provides fundamentals for further development of lignin utilization technologies.

Notation

Abbreviations

ABC	after burning chamber	H	(p–)hydroxyphenyl
ACM	Aspen Custom Modeler®	HDO	hydrodeoxygenation
BET	Brunauer-Emmett-Teller method	HDPE	high-density polyethylene
BFB	bubbling fluidized bed	HHV	higher heating value
BFD1	backwards finite difference method of first order	HL	hydrolysis lignin
BM	bed material	HPAEC	high performance anion-exchange chromatography
C1	primary cyclone	ICP	inductive coupled plasma
C2	secondary cyclone (material)	ID	inner diameter
CFB	circulating fluidized bed	KFT	Karl Fischer titration
CFBC	circulating fluidized bed combustor	KL	Kraft lignin
CFD	computational fluid dynamics	MS	mass spectrometry
CSTR	continuous flow stirred tank reactor	MWL	milled wood lignin
DAE	differential algebraic equations	NDIR	non-dispersive infrared
DCM	methylene chloride	OES	optical emission spectrometry
DMSO	dimethyl sulfoxide	PSD	particle size distribution
DR	Dubinin-Radushkevich method	S	syringyl
DSC	differential scanning calorimetry	SEC	size exclusion chromatography
EDX	energy-dispersive X-ray spectroscopy	SEM	scanning electron microscopy
FID	flame ionization detector	TCD	thermal conductivity detector
FTIR	Fourier transform infrared spectroscopy	TDH	transport disengaging height
G	guaiacyl	TG or TGA	thermogravimetry(/ic analysis)
GC	gas chromatography	UV	ultraviolet

Constants

g	gravitational acceleration (earth)	$9.81\,\mathrm{J/(mol \cdot K)}$
R	ideal gas constant	$8.314\,\mathrm{J/(mol \cdot K)}$
π	mathematical constant	3.1416

Dimensionless numbers

Ar	Archimedes number
Bi	Biot number
$D_{\mathrm{p}}^{\star}$	dimensionless particle diameter
Re	Reynolds number
$U_{\mathrm{t}}^{\star}$	dimensionless terminal velocity

Greek

ε	porosity/ volume fraction	$\mathrm{m^3/m^3}$
$\varepsilon_{\mathrm{s},\infty}$	solids volume content above TDH	$\mathrm{m^3/m^3}$
ζ	parameter for $\dot{V}_{\mathrm{b}}$	$-$
η	dynamic viscosity	$\mathrm{kg/(m \cdot s)}$
η_{OR}	oil recovery ratio	$\mathrm{kg/(m \cdot s)}$
η_{SD}	surplus-to-deficit ratio of integrated process	$\mathrm{kg/(m \cdot s)}$
θ	mechanism factor (char combustion)	$-$
ϑ	temperature	$^{\circ}\mathrm{C}$
ϑ_{g}	glass transition temperature	$^{\circ}\mathrm{C}$
ι	fit parameter	acc. to fit cor.
λ	mean duration of bubble life	s
λ_{T}	thermal conductivity	$\mathrm{W/(m \cdot K)}$
ν	stoichiometric coefficient	$-$
ξ	parameter for u_{b}	$-$
ρ	density	$\mathrm{kg/m^3}$
Σ_{v}	diffusion volume	$-$
τ	residence time	s
φ, ϕ	volume fraction	$\mathrm{m^3/m^3}$
χ	inorganics accumulation factor	$\mathrm{kg/kg}$
ψ_{Wa}	sphericity by Wadell	$-$

Latin

a	decay constant (cf. Eq. 5.22)	$1/\mathrm{m}$
A	area	$\mathrm{m^2}$
a_{b}	volume specific bubble interface area	$1/\mathrm{m}$
c	concentration	$\mathrm{mol/m^3}$
c_{p}	heat capacity	$\mathrm{kJ/(kg \cdot K)}$
C_{d}	drag coefficient	$-$
c_{vd}	solids volume concentration in dense phase	$-$
d	diameter	$\mathrm{m, \mu m}$
$D_{i,\mathrm{g}}$	binary diffusion coefficient	$\mathrm{cm^2/s}$
E	error	$\%$
E_{A}	activation energy	$\mathrm{kJ/mol}$
f	solids loading	$\mathrm{kg/kg}$
F_{d}	drag force per projection area of particle	Pa

F_g	net buoyancy force per projection area of particle	Pa
G_s	entrainment rate at reactor outlet	$kg/(m^2 \cdot s)$
$G_{s,\infty}$	entrainment rate above TDH	$kg/(m^2 \cdot s)$
h	height	m
$\dot{H}$	enthalpy flow	W
H_0	higher heating value	MJ/kg
h_T	heat transfer coefficient	$W/(m^2 \cdot K)$
$\Delta H_{R,pyr}$	pyrolysis reaction enthalpy	MJ/kg
k_0	pre-exponential factor	1/s
k	reaction rate constant	1/s
K_g	bubble-suspension mass transfer coefficient	m/s
K_q	mass exchange rate due to reaction	1/s
m	mass	kg
M	molar mass	g/mol
$\dot{m}$	mass flow	kg/s
n	amount of substance	mol
N	number	–
$\dot{n}$	molar flow	mol/s
p	pressure	Pa
P	polymerization degree	–
p^*	ratio of CO to CO_2	–
PD	polydispersity index	–
$\dot{Q}$	heat flow	W
r	reaction rate	$mol/(m^3 \cdot s)$
s	char layer thickness	µm
S	particle shrinkage	vol.-%
SR	split rate	–
t	time	s
T	temperature	K
u	velocity	m/s
u_0	superficial gas velocity	m/s
u_{solid}	mean solid velocity in dilute zone	m/s
u_t	single particle terminal velocity	m/s
V	volume	m^3
$\dot{V}$	volume flow	m^3/s
$\dot{V}_b$	visible volumetric bubble flow based on cross-sectional area of the bed	m/s
$\dot{V}_n$	volume flow through single orifice	m^2/s
w	mass fraction	g/g
X	conversion	kg/kg
$\hat{X}$	char formation mass ratio	m^3/m^3
X_{RR}	recovery rate	kg/kg
y	molar fraction	mol/mol
Y	yield	kg/kg

Subscripts

#	no. to corr. pos. in flowsheet	L	lignin
a	apparent	lit	literature
A	ash	m	material/ mean
AH	air heater	mf	minimal fluidization
amb	ambient	n	number average
b	biomass/ bubble	p	particle
BM	bed material	P	polymerization degree
C2	secondary cyclone material	PR	pyrolysis reactor
comb	combustion/ combustion reactor	PS	pyrolysis stream
db	dense bed	QS	quartz sand
exp	experiment	R	reactor/ reaction
f	fluid	rel	relative
fb	freeboard	s	solid
fbc	fluid bed cooler	Sc1	scrubber
g	gas	SS	side stream
GA	gas analyzer	susp	suspension
GB	gas bag	t	time
GH	gas heater	w	weight average
I	inert		

Superscripts

conv	convection	S	saturation
reac	reaction		

Bibliography

[1] P. Sannigrahi, Y. Pu, and A. Ragauskas. Cellulosic biorefineries—unleashing lignin opportunities. *Curr. Opin. Env. Sust.*, 2(5-6):383–393, 2010.

[2] US Energy Information Administration. Crude Oil Proved Reserves: web resource, 2016.

[3] BP p.l.c. BP Statistical Review of World Energy, 2016. URL bp.com/statisticalreview.

[4] *World population prospects: The 2015 revision*. United Nations, New York, 2015. ISBN 978-92-1-151530-5.

[5] G. Maggio and G. Cacciola. When will oil, natural gas, and coal peak? *Fuel*, 98: 111–123, 2012.

[6] F. Cherubini and A. H. Strømman. Chemicals from lignocellulosic biomass: opportunities, perspectives, and potential of biorefinery systems. *Biofuels, Bioprod. Bioref.*, 5:548–561, 2011.

[7] D. Stewart. Lignin as a base material for materials applications: Chemistry, application and economics. *Ind. Crops Prod.*, 27(2):202–207, 2008.

[8] OPEC. Global demand outlook for selected oil products worldwide from 2015 to 2040 (in million barrels per day), 2016. URL https://www.statista.com/statistics/282774/global-product-demand-outlook-worldwide/.

[9] E. Kemf, editor. *GCO–Global Chemicals Outlook: Towards sound management of chemicals*. United Nations Environment Programme, Nairobi, Kenya, 2013. ISBN 978-92-807-3320-4.

[10] S. Nonhebel. Global food supply and the impacts of increased use of biofuels: 7th Biennial International Workshop "Advances in Energy Studies". *Energy*, 37(1):115–121, 2012.

[11] R. J. Bothast and M. A. Schlicher. Biotechnological processes for conversion of corn into ethanol. *Appl. Microbiol. Biotechnol.*, 67(1):19–25, 2005.

[12] A. Pessoa-Jr, I. Conceição Roberto, M. Menossi, R. R. dos Santos, S. O. Filho, and T. C. V. Penna. Perspectives on Bioenergy and Biotechnology in Brazil. *Appl. Biochem. Biotechnol.*, 121(1-3):59–70, 2005.

[13] T. Ingram, K. Wörmeyer, J. C. I. Lima, V. Bockemühl, G. Antranikian, G. Brunner, and I. Smirnova. Comparison of different pretreatment methods for lignocellulosic materials. Part I: Conversion of rye straw to valuable products. *Bioresour. Technol.*, 102(8):5221–5228, 2011.

[14] B. Kamm, M. Kamm, M. Schmidt, T. Hirth, and M. Schulze. *Biorefineries-Industrial Processes and Products: Lignocellulose-based Chemical Products and Product Family Trees: Chapter 3*. Wiley-VCH Verlag GmbH, 2008. ISBN 9783527619849.

[15] B. Kamm, P. R. Gruber, and M. Kamm. Biorefineries – Industrial Processes and Products. In *Ullmann's Encyclopedia of Industrial Chemistry*. Wiley-VCH Verlag GmbH & Co. KGaA, Weinheim and Germany, 2000. ISBN 3527306730.

[16] V. Menon and M. Rao. Trends in bioconversion of lignocellulose: Biofuels, platform chemicals & biorefinery concept. *Prog. Energy Combust. Sci.*, 38(4):522–550, 2012.

[17] M. Alvarado-Morales, J. Terra, K. V. Gernaey, J. M. Woodley, and R. Gani. Biorefining: Computer aided tools for sustainable design and analysis of bioethanol production. *Chem. Eng. Res. Des.*, 87(9):1171–1183, 2009.

[18] S. Fahd, G. Fiorentino, S. Mellino, and S. Ulgiati. Cropping bioenergy and biomaterials in marginal land: The added value of the biorefinery concept: 7th Biennial International Workshop "Advances in Energy Studies". *Energy*, 37(1):79–93, 2012.

[19] M. G. Mandl. Status of green biorefining in Europe. *Biofuels, Bioprod. Bioref.*, 4 (3):268–274, 2010.

[20] P. J. de Wild, W. J. J. Huijgen, and R. J. A. Gosselink. Lignin pyrolysis for profitable lignocellulosic biorefineries. *Biofuels, Bioprod. Bioref.*, 8(5):645–657, 2014.

[21] C. E. Wyman. What is (and is not) vital to advancing cellulosic ethanol. *Trends Biotechnol.*, 25(4):153–157, 2007.

[22] M. Kleinert and T. Barth. Phenols from lignin. *Chem. Eng. Technol.*, 31(5):736–745, 2008.

[23] U. Schmitt, G. Koch, and R. Lehnen. Wood. In *Ullmann's Encyclopedia of Industrial Chemistry*. Wiley-VCH Verlag GmbH & Co. KGaA, Weinheim and Germany, 2000. ISBN 3527306730.

[24] M. Alexander. Biodegradation: Problems of Molecular Recalcitrance and Microbial Fallibility. In Wayne W. Umbreit, editor, *Advances in Applied Microbiology*, volume 7, pages 35–80. Academic Press, 1965. ISBN 0065-2164.

[25] A. V. Bridgwater and G. V. C. Peacocke. Fast pyrolysis processes for biomass. *Renew. Sust. Energ. Rev.*, 4(1):1–73, 2000.

[26] R. D. Perlack. Biomass as feedstock for a bioenergy and bioproducts industry: the technical feasibility of a billion-ton annual supply, 2005. URL https://www1.eere.energy.gov/bioenergy/pdfs/final_billionton_vision_report2.pdf.

[27] S.-T. Yang. Chapter 1 – Bioprocessing – from Biotechnology to Biorefinery. In S.-T. Yang, editor, *Bioprocessing for Value-Added Products from Renewable Resources*, pages 1–24. Elsevier, Amsterdam, 2007. ISBN 978-0-444-52114-9.

[28] A. U. Buranov and G. Mazza. Lignin in straw of herbaceous crops. *Ind. Crops Prod.*, 28(3):237–259, 2008.

[29] P. J. de Wild, W. J. J. Huijgen, and H. J. Heeres. Pyrolysis of wheat straw-derived organosolv lignin. *J. Anal. Appl. Pyrolysis*, 93:95–103, 2012.

[30] P. Azadi, O. R. Inderwildi, R. Farnood, and D. A. King. Liquid fuels, hydrogen and chemicals from lignin: A critical review. *Renew. Sust. Energ. Rev.*, 21:506–523, 2013.

[31] B. Saake and R. Lehnen. Lignin. In *Ullmann's Encyclopedia of Industrial Chemistry*. Wiley-VCH Verlag GmbH & Co. KGaA, Weinheim and Germany, 2000. ISBN 3527306730.

[32] M. Ragnar, G. Henriksson, M. E. Lindström, M. Wimby, and R. Süttinger. Pulp. In *Ullmann's Encyclopedia of Industrial Chemistry*, pages 1–89. Wiley-VCH Verlag GmbH & Co. KGaA, Weinheim and Germany, 2000. ISBN 3527306730.

[33] A. Limayem and S. C. Ricke. Lignocellulosic biomass for bioethanol production: Current perspectives, potential issues and future prospects. *Prog. Energy Combust. Sci.*, 38(4):449–467, 2012.

[34] F. Shafizadeh, R. H. Furneaux, T. G. Cochran, J. P. Scholl, and Y. Sakai. Production of levoglucosan and glucose from pyrolysis of cellulosic materials. *J. Appl. Polym. Sci.*, 23(12):3525–3539, 1979.

[35] H. Krässig, J. Schurz, R. G. Steadman, K. Schliefer, W. Albrecht, M. Mohring, and H. Schlosser. Cellulose. In *Ullmann's Encyclopedia of Industrial Chemistry*. Wiley-VCH Verlag GmbH & Co. KGaA, Weinheim and Germany, 2000. ISBN 3527306730.

[36] D. Mohan, C. U. Pittman, and P. H. Steele. Pyrolysis of wood/biomass for bio-oil: A critical review. *Energy & Fuels*, 20(3):848–889, 2006.

[37] *Römpp Chemie Lexikon: CD Version 2.0*. Georg Thieme Verlag, Stuttgart/ New York, 1999.

[38] R. F. H. Dekker. Biodegradation of the Hetero-1,4-Linked Xylans. In *Plant Cell Wall Polymers*, volume 399 of *ACS Symposium Series*, pages 619–629. American Chemical Society, 1989. ISBN 0-8412-1658-4.

[39] K. Poutanen, M. Sundberg, H. Korte, and J. Puls. Deacetylation of xylans by acetyl esterases of Trichoderma reesei. *Applied Microbiology and Biotechnology*, 33 (5):506–510, 1990.

[40] de Candolle, M. A. P. *Théorie Élémentaire de la Botanique*. Déterville, Paris, 1813.

[41] K. Wörmeyer, T. Ingram, B. Saake, G. Brunner, and I. Smirnova. Comparison of different pretreatment methods for lignocellulosic materials. Part II: Influence of pretreatment on the properties of rye straw lignin. *Bioresour. Technol.*, 102(5): 4157–4164, 2011.

[42] C. Amen-Chen, H. Pakdel, and C. Roy. Production of monomeric phenols by thermochemical conversion of biomass: a review. *Bioresour. Technol.*, 79(3):277–299, 2001.

[43] Pandey, M. P. and Kim, C. S. Lignin Depolymerization and Conversion: A Review of Thermochemical Methods. *Chem. Eng. Technol.*, 34(1):29–41, 2011.

[44] A. Tolbert, H. Akinosho, R. Khunsupat, A. K. Naskar, and A. J. Ragauskas. Characterization and analysis of the molecular weight of lignin for biorefining studies. *Biofuels, Bioprod. Bioref.*, 8(6):836–856, 2014.

[45] A. G. Gayubo, B. Valle, A. T. Aguayo, M. Olazar, and J. Bilbao. Pyrolytic lignin removal for the valorization of biomass pyrolysis crude bio-oil by catalytic transformation. *J. Chem. Technol. Biotechnol.*, 85(1):132–144, 2010.

[46] R. Bayerbach. *Über die Struktur der oligomeren Bestandteile von Flash-Pyrolyseölen aus Biomasse.* PhD thesis, Universität Hamburg, Hamburg, 2006.

[47] I. Brodin, E. Sjöholm, and G. Gellerstedt. The behavior of kraft lignin during thermal treatment. *J. Anal. Appl. Pyrolysis*, 87(1):70–77, 2010.

[48] D. J. Nowakowski, A. V. Bridgwater, D. C. Elliott, D. Meier, and P. de Wild. Lignin fast pyrolysis: Results from an international collaboration. *J. Anal. Appl. Pyrolysis*, 88(1):53–72, 2010.

[49] T. Hatakeyama, K. Nakamura, and H. Hatakeyama. Studies on heat capacity of cellulose and lignin by differential scanning calorimetry. *Polymer*, 23(12):1801–1804, 1982.

[50] Z. Y. Luo, S. R. Wang, and K. F. Cen. A model of wood flash pyrolysis in fluidized bed reactor. *Renewable Energy*, 30(3):377–392, 2005.

[51] D. S. Scott and J. Piskorz. The flash pyrolysis of aspen-poplar wood. *Can. J. Chem. Eng.*, 60(5):666–674, 1982.

[52] M. Amutio, G. Lopez, R. Aguado, M. Artetxe, J. Bilbao, and M. Olazar. Effect of Vacuum on Lignocellulosic Biomass Flash Pyrolysis in a Conical Spouted Bed Reactor: Energy & Fuels. *Energy & Fuels*, 25(9):3950–3960, 2011.

[53] A. V. Bridgwater. Principles and practice of biomass fast pyrolysis processes for liquids. *J. Anal. Appl. Pyrolysis*, 51(1-2):3–22, 1999.

[54] A. V. Bridgwater, D. Meier, and D. Radlein. An overview of fast pyrolysis of biomass. *Organic Geochemistry*, 30(12):1479–1493, 1999.

[55] A. V. Bridgwater. Renewable fuels and chemicals by thermal processing of biomass. *Chem. Eng. J.*, 91(2-3):87–102, 2003.

[56] A. V. Bridgwater. Review of fast pyrolysis of biomass and product upgrading. *Biomass and Bioenergy*, 38:68–94, 2012.

[57] T. R. Brown, M. M. Wright, and R. C. Brown. Estimating profitability of two biochar production scenarios: slow pyrolysis vs fast pyrolysis. *Biofuels, Bioprod. Bioref.*, 5(1):54–68, 2011.

[58] B.-S. Kang, K. H. Lee, H. J. Park, Y.-K. Park, and J. S. Kim. Fast pyrolysis of radiata pine in a bench scale plant with a fluidized bed: Influence of a char separation system and reaction conditions on the production of bio-oil. *J. Anal. Appl. Pyrolysis*, 76(1-2):32–37, 2006.

[59] N. Abatzoglou, M. Gagnon, and E. Chornet. Hot gas filtration via a novel mobile granular filter. In A. V. Bridgwater, editor, *Progress in thermochemical biomass conversion*, volume 1, pages 365–378. Blackwell Science, Oxford, 2001. ISBN 0-632-05533-2.

[60] R. C. Brown, J. Smeenk, and C. Wistrom. Design of a moving bed granular filter for biomass gasification. In A. V. Bridgwater, editor, *Progress in thermochemical biomass conversion*, volume 1, pages 379–387. Blackwell Science, Oxford, 2001. ISBN 0-632-05533-2.

[61] C. Paenpong, S. Inthidech, and A. Pattiya. Effect of filter media size, mass flow rate and filtration stage number in a moving-bed granular filter on the yield and properties of bio-oil from fast pyrolysis of biomass. *Bioresour. Technol.*, 139:34–42, 2013.

[62] C. Paenpong and A. Pattiya. Filtration of fast pyrolysis char fines with a cross-flow moving-bed granular filter. *Powder Technol.*, 245:233–240, 2013.

[63] G. V. C. Peacocke, C. M. Dick, R. A. Hague, L. A. Cooke, and A. V. Bridgwater. Comparison of ablative and fluid bed fast pyrolysis products: yields and analyses: Biomass fast pyrolysis for liquids. In A. V. Bridgwater and D. G. B. Boocock, editors, *Developments in thermochemical biomass conversion*, pages 191–205. Blackie Academic & Professional, London and New York, 1997. ISBN 0751403504.

[64] G. Brem. *Flash-Pyrolysis in a cyclone*. Patent, US 7,202,389 B1, 2007.

[65] R. Galiasso, Y. González, and M. Lucena. New inverted cyclone reactor for flash hydropyrolysis. *Int. Symp. on Adv. in Hydroprocessing of Oil Fractions*, 220–222:186–197, 2014.

[66] J. Lédé, F. Broust, F.-T. Ndiaye, and M. Ferrer. Properties of bio-oils produced by biomass fast pyrolysis in a cyclone reactor. *Fuel*, 86(12–13):1800–1810, 2007.

[67] J. Lédé. The Cyclone: A Multifunctional Reactor for the Fast Pyrolysis of Biomass. *Ind. Eng. Chem. Res*, 39(4):893–903, 2000.

[68] H. Wiinikka, P. Carlsson, A.-C. Johansson, M. Gullberg, C. Ylipää, M. Lundgren, L. Sandström, and M. Lundgren. Fast Pyrolysis of Stem Wood in a Pilot-Scale Cyclone Reactor. *Energy & Fuels*, 29(5):3158–3167, 2015.

[69] R. S. Miller and J. Bellan. Numerical Simulation of Vortex Pyrolysis Reactors for Condensable Tar Production from Biomass. *Energy & Fuels*, 12(1):25–40, 1998.

[70] H. Hartmann and M. Kaltschmitt. *Energie aus Biomasse*. Springer Berlin Heidelberg, Berlin and Heidelberg, 2009. ISBN 978-3-540-85095-3.

[71] J. P. Diebold and A. Power. Engineering Aspects of the Vortex Pyrolysis Reactor to Produce Primary Pyrolysis Oil Vapors for Use in Resins and Adhesives. In A. V. Bridgwater and J. L. Kuester, editors, *Research in thermochemical biomass conversion*, pages 609–628. Elsevier Applied Science and Sole distributor in the USA and Canada, Elsevier Science Pub. Co, London and New York and New York, NY, USA, 1988. ISBN 9781851663101.

[72] R. W. Ashcraft, G. J. Heynderickx, and G. B. Marin. Modeling fast biomass pyrolysis in a gas–solid vortex reactor: 22nd International Symposium on Chemical Reaction Engineering (ISCRE 22). *Chem. Eng. J.*, 207–208:195–208, 2012.

[73] N. Dahmen and E. Dinjus. Synthetische Chemieprodukte und Kraftstoffe aus Biomasse. *Chem. Ing. Tech*, 82(8):1147–1152, 2010.

[74] P. Kim, A. Johnson, C. W. Edmunds, M. Radosevich, F. Vogt, T. G. Rials, and N. Labbé. Surface Functionality and Carbon Structures in Lignocellulosic-Derived Biochars Produced by Fast Pyrolysis. *Energy & Fuels*, 25(10):4693–4703, 2011.

[75] A. M. C. Janse, W. Prins, and W. P. M. van Swaaij. Development of a small integrated pilot plant for flash pyrolysis of biomass. In A. V. Bridgwater and D. G. B. Boocock, editors, *Developments in thermochemical biomass conversion*, pages 368–377. Blackie Academic & Professional, London and New York, 1997. ISBN 0751403504.

[76] J. Lian, M. Garcia-Perez, R. Coates, H. W. Wu, and S. Chen. Yeast fermentation of carboxylic acids obtained from pyrolytic aqueous phases for lipid production. *Bioresour. Technol.*, 118:177–186, 2012.

[77] J. Yang, D. Blanchette, B. de Caumina, and C. Roy. Modelling, Scale-up and Demonstration of a Vacuum Pyrolysis Reactor. In A. V. Bridgwater, editor, *Progress in thermochemical biomass conversion*, volume 2, pages 1296–1311. Blackwell Science, Oxford, 2001. ISBN 0-632-05533-2.

[78] B. Maddi, S. Viamajala, and S. Varanasi. Comparative study of pyrolysis of algal biomass from natural lake blooms with lignocellulosic biomass. *Bioresour. Technol.*, 102(23):11018–11026, 2011.

[79] O. Onay and O. M. Koçkar. Fixed-bed pyrolysis of rapeseed (Brassica napus L.). *Biomass and Bioenergy*, 26(3):289–299, 2004.

[80] G. Elordi, M. Olazar, G. Lopez, M. Artetxe, and J. Bilbao. Product Yields and Compositions in the Continuous Pyrolysis of High-Density Polyethylene in a Conical Spouted Bed Reactor: Industrial & Engineering Chemistry Research. *Ind. Eng. Chem. Res*, 50(11):6650–6659, 2011.

[81] M. Amutio, G. Lopez, R. Aguado, J. Bilbao, and M. Olazar. Biomass Oxidative Flash Pyrolysis: Autothermal Operation, Yields and Product Properties: Energy & Fuels. *Energy & Fuels*, 26(2):1353–1362, 2012.

[82] A. Atutxa, R. Aguado, A. G. Gayubo, M. Olazar, and J. Bilbao. Kinetic Description of the Catalytic Pyrolysis of Biomass in a Conical Spouted Bed Reactor. *Energy & Fuels*, 19(3):765–774, 2005.

[83] D. S. Scott and J. Piskorz. The flash pyrolysis of aspen-poplar wood. In E. P. O. National Research Council of Canada, editor, *Proceedings, third Bioenergy R&D Seminar*, pages 265–272, 1981.

[84] D. S. Scott and J. Piskorz. The continuous flash pyrolysis of biomass. *Can. J. Chem. Eng.*, 62(3):404–412, 1984.

[85] V. Lago. *Application of Mechanically Fluidized Reactors to Lignin Pyrolysis*. PhD thesis, Western University, London, Ontario, Canada, April 2015.

[86] V. Lago, C. Greenhalf, C. Briens, and F. Berruti. Mixing and operability characteristics of mechanically fluidized reactors for the pyrolysis of biomass. *Powder Technol.*, 274:205–212, 2015.

[87] A. Tumbalam Gooty, D. Li, F. Berruti, and C. Briens. Kraft-lignin pyrolysis and fractional condensation of its bio-oil vapors. *J. Anal. Appl. Pyrolysis*, 106:33–40, 2014.

[88] I. P. Boukis, P. Grammelis, S. Bezergianni, and A. V. Bridgwater. CFB air-blown flash pyrolysis. Part I: Engineering design and cold model performance. *Fuel*, 86 (10-11):1372–1386, 2007.

[89] I. P. Boukis, S. Bezergianni, P. Grammelis, and A. V. Bridgwater. CFB air-blown flash pyrolysis. Part II: Operation and experimental results. *Fuel*, 86(10-11):1387–1395, 2007.

[90] A. Lappas, M. Samolada, D. Iatridis, S. Voutetakis, and I. Vasalos. Biomass pyrolysis in a circulating fluid bed reactor for the production of fuels and chemicals. *Fuel*, 81 (16):2087–2095, 2002.

[91] M. van de Velden, J. Baeyens, and L. Boukis. Modeling CFB biomass pyrolysis reactors. *Biomass and Bioenergy*, 32(2):128–139, 2008.

[92] M. Franck, E.-U. Hartge, S. Heinrich, B. Lorenz, and J. Werther. Energetic Optimization of the Lignin Pyrolysis for the Production of Aromatic Hydrocarbons. In T. M. Knowlton, editor, *Proceedings of the Tenth International Conference on Circulating Fluidized Beds and Fluidization Technology - CFB-10*, pages 241–248. Engineering Conferences International, 2011. ISBN 978-1-4507-7082-5.

[93] M. Franck, E.-U. Hartge, S. Heinrich, J. Werther, and D. Meier. Lignin Pyrolysis in a Circulating Fluidized Bed – Influence of Temperature and Char Formation on Bed Material. In J. Li, F. Wei, X. Bao, and W. Wang, editors, *Proc. 11th Int. Conf. on Fluidized Bed Technol.*, volume 1, pages 733–738. Chemical Industry Press, 2014. ISBN 978-7-122-20169-0.

[94] A. L. Brown, D. C. Dayton, M. R. Nimlos, and J. W. Daily. Design and Characterization of an Entrained Flow Reactor for the Study of Biomass Pyrolysis Chemistry at High Heating Rates. *Energy & Fuels*, 15(5):1276–1285, 2001.

[95] J. M. Commandré, H. Lahmidi, S. Salvador, and N. Dupassieux. Pyrolysis of wood at high temperature: The influence of experimental parameters on gaseous products. *Fuel Process. Technol.*, 92(5):837–844, 2011.

[96] C. Gustafsson and T. Richards. Pyrolysis kinetics of washed precipitated lignin. *Bioresources*, 4(1):26–37, 2009.

[97] D. J. Macquarrie, J. H. Clark, and E. Fitzpatrick. The microwave pyrolysis of biomass. *Biofuels, Bioprod. Bioref.*, page n/a, 2012.

[98] C. Yin. Microwave-assisted pyrolysis of biomass for liquid biofuels production. *Bioresour. Technol.*, 120:273–284, 2012.

[99] Z. Abubakar, A. A. Salema, and F. N. Ani. A new technique to pyrolyse biomass in a microwave system: Effect of stirrer speed. *Bioresour. Technol.*, 128:578–585, 2013.

[100] L. Burhenne, J. Messmer, T. Aicher, and M.-P. Laborie. The effect of the biomass components lignin, cellulose and hemicellulose on TGA and fixed bed pyrolysis. *J. Anal. Appl. Pyrolysis*, 101:177–184, 2013.

[101] S. Wang, X. Guo, K. Wang, and Z. Y. Luo. Influence of the interaction of components on the pyrolysis behavior of biomass. *J. Anal. Appl. Pyrolysis*, 91(1):183–189, 2011.

[102] T. Qu, W. Guo, L. Shen, J. Xiao, and K. Zhao. Experimental Study of Biomass Pyrolysis Based on Three Major Components: Hemicellulose, Cellulose, and Lignin: Industrial & Engineering Chemistry Research. *Ind. Eng. Chem. Res*, 50(18):10424–10433, 2011.

[103] M. Windt, D. Meier, J. H. Marsman, H. J. Heeres, and S. de Koning. Micro-pyrolysis of technical lignins in a new modular rig and product analysis by GC–MS/FID and GC×GC–TOFMS/FID. *J. Anal. Appl. Pyrolysis*, 85(1-2):38–46, 2009.

[104] H. S. Choi, D. Meier, and M. Windt. Rapid screening of catalytic pyrolysis reactions of Organosolv lignins with the vTI-mini fast pyrolyzer. *Environ. Prog. Sustain. Energy*, 31(2):240–244, 2012.

[105] J. I. Montoya, C. Valdés, F. Chejne, C. A. Gómez, A. Blanco, G. Marrugo, J. Osorio, E. Castillo, J. Aristóbulo, and J. Acero. Bio-oil production from Colombian bagasse by fast pyrolysis in a fluidized bed: An experimental study. *J. Anal. Appl. Pyrolysis*, 112:379–387, 2015.

[106] J. Werther. Fluidized-Bed Reactors. *Ullmann's Encyclopedia of Industrial Chemistry*, 2000.

[107] B. A. Freel, D. Clarke, and J. F. Kriz. *Rapid thermal processing of heavy hydrocarbon feedstocks in the presence of calcium*. Patent, EP 1 420 058 A1, 2004.

[108] B. A. Freel and R. G. Graham. *Apparatus for a circulating bed transport fast pyrolysis reactor system*. Patent, US000005961786A, 1999.

[109] B. A. Freel and R. G. Graham. *Method and Apparatus for a Circulating Bed Transport Fast Pyrolysis Reactor System*. Patent, CA000002009021C, 2001.

[110] R. G. Graham and B. A. Freel. *Products produced from rapid thermal processing of heavy hydrocarbon feedstock*. Patent, CA000002422534A1, 28.03.2002.

[111] Y. Solantausta, A. Oasmaa, K. Sipilä, C. Lindfors, J. Lehto, J. Autio, P. Jokela, J. Alin, and J. Heiskanen. Bio-oil Production from Biomass: Steps toward Demonstration: Energy & Fuels. *Energy & Fuels*, 26(1):233–240, 2011.

[112] D. Meier, B. van de Beld, A. V. Bridgwater, D. C. Elliott, A. Oasmaa, and F. Preto. State-of-the-art of fast pyrolysis in IEA bioenergy member countries. *Renew. Sust. Energ. Rev.*, 20:619–641, 2013.

[113] Q. Yang, S. B. Wu, R. Lou, and G. J. Lv. Analysis of wheat straw lignin by thermogravimetry and pyrolysis-gas chromatography/mass spectrometry. *J. Anal. Appl. Pyrolysis*, 87(1):65–69, 2010.

[114] J. A. Caballero, R. Font, and A. Marcilla. Study of the primary pyrolysis of Kraft lignin at high heating rates: yields and kinetics. *J. Anal. Appl. Pyrolysis*, 36(2):159–178, 1996.

[115] D. Ferdous, A. K. Dalai, S. K. Bej, and R. W. Thring. Production of hydrogen and medium BTU gas via pyrolysis of a kraft lignin in a fixed-bed reactor. *Proc. Intersoc. Energy Convers. Eng. Conf.*, 2:782–792, 2000.

[116] H. S. Choi and D. Meier. Fast pyrolysis of Kraft lignin—Vapor cracking over various fixed-bed catalysts. *J. Anal. Appl. Pyrolysis*, 100:207–212, 2013.

[117] D. Ferdous, A. K. Dalai, S. K. Bej, R. W. Thring, and N. N. Bakhshi. Production of H2 and medium Btu gas via pyrolysis of lignins in a fixed-bed reactor. *Fuel Process. Technol.*, 70(1):9–26, 2001.

[118] R. Lou, S. B. Wu, and G. J. Lv. Effect of conditions on fast pyrolysis of bamboo lignin. *J. Anal. Appl. Pyrolysis*, 89(2):191–196, 2010.

[119] E. Avni, F. Davoudzadeh, and R. W. Coughlin. Flash Pyrolysis of Lignin. In R. Overend, T. Milne, and L. Mudge, editors, *Fundamentals of thermochemical biomass conversion*, pages 329–344. Elsevier Applied Science Publishers, London, New York, 1985. ISBN 0-85334-306-3.

[120] Y. Huang, Z. Wei, Z. Qiu, X. Yin, and C. Wu. Study on structure and pyrolysis behavior of lignin derived from corncob acid hydrolysis residue. *J. Anal. Appl. Pyrolysis*, 93:153–159, 2012.

[121] H. E. Jegers and M. T. Klein. Primary and secondary lignin pyrolysis reaction pathways. *Ind. Eng. Chem. Proc. Des. Dev.*, 24(1):173–183, 1985.

[122] L. Qian, S. Wang, Y. Zheng, Z. Y. Luo, and K. Cen. Mechanism study of wood lignin pyrolysis by using TG-FTIR analysis. *J. Anal. Appl. Pyrolysis*, 82(1):170–177, 2008.

[123] R. K. Sharma, J. B. Wooten, V. L. Baliga, X. H. Lin, W. G. Chan, and M. R. Hajaligol. Characterization of chars from pyrolysis of lignin. *Fuel*, 83(11-12):1469–1482, 2004.

[124] T. N. Trinh, P. A. Jensen, Z. Sárossy, K. Dam-Johansen, N. O. Knudsen, H. R. Sørensen, and H. Egsgaard. Fast Pyrolysis of Lignin Using a Pyrolysis Centrifuge Reactor. *Energy & Fuels*, 27(7):3802–3810, 2013.

[125] S. H. Beis, S. Mukkamala, N. Hill, J. Joseph, C. Baker, B. Jensen, E. A. Stemmler, M. C. Wheeler, B. G. Frederick, A. van Keiningen, A. G. Berg, and W. J. DeSisto. Fast Pyrolysis of Lignins. *Bioresources*, 5(3):1408–1424, 2010.

[126] T. R. Nunn, J. B. Howard, J. P. Longwell, and W. A. Peters. Product compositions and kinetics in the rapid pyrolysis of milled wood lignin. *Ind. Eng. Chem. Proc. Des. Dev.*, 24(3):844–852, 1985.

[127] H. Ben and A. J. Ragauskas. NMR Characterization of Pyrolysis Oils from Kraft Lignin. *Energy & Fuels*, 25(5):2322–2332, 2011.

[128] E. Salehi, J. Abedi, and T. G. Harding. Bio-oil from Sawdust: Effect of Operating Parameters on the Yield and Quality of Pyrolysis Products. *Energy & Fuels*, page 110826080828063, 2011.

[129] E. Hoekstra, W. P. M. van Swaaij, S. R. A. Kersten, and K. J. A. Hogendoorn. Fast pyrolysis in a novel wire-mesh reactor: Decomposition of pine wood and model compounds. *Chem. Eng. J.*, 187:172–184, 2012.

[130] Y. Chhiti, S. Salvador, J.-M. Commandré, and F. Broust. Thermal decomposition of bio-oil: Focus on the products yields under different pyrolysis conditions: Special Section: ACS Clean Coal. *Fuel*, 102:274–281, 2012.

[131] C. Di Blasi. Modelling the fast pyrolysis of cellulosic particles in fluid-bed reactors. *Chem. Eng. Sci.*, 55(24):5999–6013, 2000.

[132] S. R. A. Kersten, X. Wang, W. Prins, and W. P. M. van Swaaij. Biomass Pyrolysis in a Fluidized Bed Reactor. Part 1: Literature Review and Model Simulations. *Ind. Eng. Chem. Res*, 44(23):8773–8785, 2005.

[133] S. N. Xiu, Z. H. Li, B. M. Li, W. M. Yi, and X. Y. Bai. Devolatilization characteristics of biomass at flash heating rate. *Fuel*, 85(5-6):664–670, 2006.

[134] J. Lédé and O. Authier. Temperature and heating rate of solid particles undergoing a thermal decomposition. Which criteria for characterizing fast pyrolysis? *J. Anal. Appl. Pyrolysis*, 113:1–14, 2015.

[135] H. J. Park, Y.-K. Park, and J. S. Kim. Influence of reaction conditions and the char separation system on the production of bio-oil from radiata pine sawdust by fast pyrolysis. *Fuel Process. Technol.*, 89(8):797–802, 2008.

[136] L. G. Wei, S. P. Xu, L. Zhang, H. G. Zhang, C. H. Liu, H. Zhu, and S. Q. Liu. Characteristics of fast pyrolysis of biomass in a free fall reactor. *Fuel Process. Technol.*, 87(10):863–871, 2006.

[137] A. E. Pütün, E. Apaydın, and E. Pütün. Rice straw as a bio-oil source via pyrolysis and steam pyrolysis. *Efficiency, Costs, Optimization, Simulation and Environmental Impact of Energy Systems*, 29(12–15):2171–2180, 2004.

[138] J. Shen, X.-S. Wang, M. Garcia-Perez, D. Mourant, M. J. Rhodes, and C.-Z. Li. Effects of particle size on the fast pyrolysis of oil mallee woody biomass. *Fuel*, 88 (10):1810–1817, 2009.

[139] M. E. Boucher, A. Chaala, and C. Roy. Bio-oils obtained by vacuum pyrolysis of softwood bark as a liquid fuel for gas turbines. Part I: Properties of bio-oil and its blends with methanol and a pyrolytic aqueous phase. *Biomass and Bioenergy*, 19 (5):337–350, 2000.

[140] P. F. Britt, Buchanan, A. C., M. J. Cooney, and D. R. Martineau. Flash Vacuum Pyrolysis of Methoxy-Substituted Lignin Model Compounds. *J. Org. Chem.*, 65(5): 1376–1389, 2000.

[141] C. Roy, A. Chaala, and H. Darmstadt. The vacuum pyrolysis of used tires: End-uses for oil and carbon black products. *J. Anal. Appl. Pyrolysis*, 51(1–2):201–221, 1999.

[142] D. S. Scott, P. Majerski, J. Piskorz, and D. Radlein. A second look at fast pyrolysis of biomass–the RTI process. *J. Anal. Appl. Pyrolysis*, 51(1-2):23–37, 1999.

[143] S. M. Ward and J. Braslaw. Experimental weight loss kinetics of wood pyrolysis under vacuum. *Combust. Flame*, 61(3):261–269, 1985.

[144] J. Lédé and O. Authier. Characterization of biomass fast pyrolysis. *Biomass Conversion and Biorefinery*, 1(3):133–147, 2011.

[145] R. Ragucci, P. Giudicianni, and A. Cavaliere. Cellulose slow pyrolysis products in a pressurized steam flow reactor. *Fuel*, 107:122–130, 2013.

[146] P. Giudicianni, G. Cardone, and R. Ragucci. Cellulose, hemicellulose and lignin slow steam pyrolysis: Thermal decomposition of biomass components mixtures. *J. Anal. Appl. Pyrolysis*, 100(0):213–222, 2013.

[147] V. Minkova, M. Razvigorova, M. Goranova, L. Ljutzkanov, and G. Angelova. Effect of Water-Vapour on the Pyrolysis of Solid Fuels: 1. Effect of Water-Vapour during the Pyrolysis of Solid Fuels on the Yield and Composition of the Liquid Products. *Fuel*, 70(6):713–719, 1991.

[148] E. P. Önal, B. B. Uzun, and A. E. Pütün. Steam pyrolysis of an industrial waste for bio-oil production. *Fuel Process. Technol.*, 92(5):879–885, 2011.

[149] V. Minkova, S. P. Marinov, R. Zanzi, E. Björnbom, T. Budinova, M. Stefanova, and L. Lakov. Thermochemical treatment of biomass in a flow of steam or in a mixture of steam and carbon dioxide. *Fuel Process. Technol.*, 62(1):45–52, 2000.

[150] V. Minkova, M. Razvigorova, E. Björnbom, R. Zanzi, T. Budinova, and N. Petrov. Effect of water vapour and biomass nature on the yield and quality of the pyrolysis products from biomass. *Fuel Process. Technol.*, 70(1):53–61, 2001.

[151] P. Mellin, E. Kantarelis, C. Zhou, and W. Yang. Simulation of Bed Dynamics and Primary Products from Fast Pyrolysis of Biomass: Steam Compared to Nitrogen as a Fluidizing Agent. *Ind. Eng. Chem. Res*, 2014.

[152] K. El harfi, A. Mokhlisse, and M. B. Chanâa. Effect of water vapor on the pyrolysis of the Moroccan (Tarfaya) oil shale. *J. Anal. Appl. Pyrolysis*, 48(2):65–76, 1999.

[153] E. Kantarelis, W. Yang, and W. Blasiak. Production of Liquid Feedstock from Biomass via Steam Pyrolysis in a Fluidized Bed Reactor. *Energy & Fuels*, 27(8): 4748–4759, 2013.

[154] R. A. Graff and S. D. Brandes. Coal Liquefaction by Steam Pyrolysis. *Am. Chem. SocDiv. Fuel Chem. Prepr.*, 29(2):104–111, 1984.

[155] J. S. Abichandani, C. Deradourian, R. E. Gannon, D. B. Stickler, J. A. Woodroffe, and K. G. Neoh. Pressure effects on steam pyrolysis of coal. *Fuel Process. Technol.*, 18(2):133–146, 1988.

[156] H. Zhang, R. Xiao, D. Wang, G. He, S. Shao, J. Zhang, and Z. Zhong. Biomass fast pyrolysis in a fluidized bed reactor under N2, CO2, CO, CH4 and H2 atmospheres. *Bioresour. Technol.*, 102(5):4258–4264, 2011.

[157] S. Thangalazhy-Gopakumar, S. Adhikari, R. B. Gupta, M. Tu, and S. Taylor. Production of hydrocarbon fuels from biomass using catalytic pyrolysis under helium and hydrogen environments. *Bioresour. Technol.*, 102(12):6742–6749, 2011.

[158] S. Meesuk, J.-P. Cao, K. Sato, Y. Ogawa, and T. Takarada. Fast Pyrolysis of Rice Husk in a Fluidized Bed: Effects of the Gas Atmosphere and Catalyst on Bio-oil with a Relatively Low Content of Oxygen. *Energy & Fuels*, page 110812134919086, 2011.

[159] M. Zhong, Z. Zhang, Q. Zhou, J. Yue, S. Gao, and G. Xu. Continuous high-temperature fluidized bed pyrolysis of coal in complex atmospheres: Product distribution and pyrolysis gas. *J. Anal. Appl. Pyrolysis*, 97:123–129, 2012.

[160] C. Jindarom, V. Meeyoo, T. Rirksomboon, and P. Rangsunvigit. Thermochemical decomposition of sewage sludge in CO2 and N2 atmosphere. *Chemosphere*, 67(8): 1477–1484, 2007.

[161] K. Raveendran, A. Ganesh, and K. C. Khilar. Influence of mineral matter on biomass pyrolysis characteristics. *Fuel*, 74(12):1812–1822, 1995.

[162] H. B. Goyal, D. Seal, and R. C. Saxena. Bio-fuels from thermochemical conversion of renewable resources: A review. *Renew. Sust. Energ. Rev.*, 12(2):504–517, 2008.

[163] T. N. Trinh, P. A. Jensen, K. Dam-Johansen, N. O. Knudsen, H. R. Sørensen, and S. Hvilsted. Comparison of Lignin, Macroalgae, Wood, and Straw Fast Pyrolysis. *Energy & Fuels*, 2013.

[164] G. N. Richards and G. Zheng. Influence of metal ions and of salts on products from pyrolysis of wood: Applications to thermochemical processing of newsprint and biomass. *J. Anal. Appl. Pyrolysis*, 21(1-2):133–146, 1991.

[165] R. Fahmi, A. V. Bridgwater, I. Donnison, N. Yates, and J. M. Jones. The effect of lignin and inorganic species in biomass on pyrolysis oil yields, quality and stability. *Fuel*, 87(7):1230–1240, 2008.

[166] I.-Y. Eom, J.-Y. Kim, S.-M. Lee, T.-S. Cho, H. Yeo, and J.-W. Choi. Comparison of pyrolytic products produced from inorganic-rich and demineralized rice straw (Oryza sativa L.) by fluidized bed pyrolyzer for future biorefinery approach. *Bioresour. Technol.*, 128:664–672, 2013.

[167] D. J. Nowakowski, J. M. Jones, R. M. D. Brydson, and A. B. Ross. Potassium catalysis in the pyrolysis behaviour of short rotation willow coppice. *Fuel*, 86(15): 2389–2402, 2007.

[168] C. Liu, X. Liu, X. Bi, Y. Liu, and C. Wang. Influence of Inorganic Additives on Pyrolysis of Pine Bark. *Energy & Fuels*, 25(5):1996–2003, 2011.

[169] A. Demirbas. Analysis of Liquid Products from Biomass via Flash Pyrolysis. *Energy Sources, Part A: Recovery, Utilization, and Environmental Effects*, 24(4):337–345, 2002.

[170] W.-P. Pan and G. N. Richards. Influence of metal ions on volatile products of pyrolysis of wood. *J. Anal. Appl. Pyrolysis*, 16(2):117–126, 1989.

[171] I.-Y. Eom, J.-Y. Kim, T.-S. Kim, S.-M. Lee, D. Choi, I.-G. Choi, and J.-W. Choi. Effect of essential inorganic metals on primary thermal degradation of lignocellulosic biomass. *Bioresour. Technol.*, 104:687–694, 2012.

[172] J. Piskorz, D. S. Radlein, D. S. Scott, and S. Czernik. Pretreatment of wood and cellulose for production of sugars by fast pyrolysis. *J. Anal. Appl. Pyrolysis*, 16(2): 127–142, 1989.

[173] D. S. Scott, L. Paterson, J. Piskorz, and D. Radlein. Pretreatment of poplar wood for fast pyrolysis: rate of cation removal. *J. Anal. Appl. Pyrolysis*, 57(2):169–176, 2001.

[174] M. R. Gray. *The effects of moisture and ash content on the pyrolysis of a wood derived material*. PhD thesis, Cal. Inst. of Tech., Pasadena, California, 1984.

[175] M. R. Gray, W. H. Corcoran, and G. R. Gavalas. Pyrolysis of a wood-derived material. Effects of moisture and ash content: Industrial & Engineering Chemistry Process Design and Development. *Ind. Eng. Chem. Proc. Des. Dev.*, 24(3):646–651, 1985.

[176] G. Dobele, G. Rossinskaja, T. Dizhbite, G. Telysheva, D. Meier, and O. Faix. Application of catalysts for obtaining 1,6-anhydrosaccharides from cellulose and wood by fast pyrolysis. *Pyrolysis 2004*, 74(1–2):401–405, 2005.

[177] P. T. Williams and P. A. Horne. The role of metal salts in the pyrolysis of biomass. *Renewable Energy*, 4(1):1–13, 1994.

[178] G. Varhegyi, M. J. Antal, E. Jakab, and P. Szabo. Kinetic modeling of biomass pyrolysis. *J. Anal. Appl. Pyrolysis*, 42(1):73–87, 1997.

[179] E. Jakab, O. Faix, F. Till, and T. Székely. The effect of cations on the thermal decomposition of lignins. *J. Anal. Appl. Pyrolysis*, 25:185–194, 1993.

[180] A. D. Paulsen, M. S. Mettler, D. G. Vlachos, and P. J. Dauenhauer. The Role of Sample Dimension and Temperature in Cellulose Pyrolysis. *Energy & Fuels*, 27(4): 2126–2134, 2013.

[181] D.-l. Guo, S. B. Wu, R. Lou, X.-l. Yin, and Q. Yang. Effect of organic bound Na groups on pyrolysis and CO2-gasification of alkali lignin. *Bioresources*, 6(4): 4145–4157, 2011.

[182] P. Rutkowski. Pyrolysis of cellulose, xylan and lignin with the K2CO3 and ZnCl2 addition for bio-oil production. *Fuel Process. Technol.*, 92(3):517–522, 2011.

[183] E. Jakab, O. Faix, and F. Till. Thermal decomposition of milled wood lignins studied by thermogravimetry/mass spectrometry. *J. Anal. Appl. Pyrolysis*, 40-41:171–186, 1997.

[184] E. Kojima, Y. Miao, and S. Yoshizaki. Pyrolysis of cellulose particles in a fluidized bed. *J. Chem. Eng. Japan*, 24(1):8–14, 1991.

[185] A. V. Bridgwater, S. Czernik, and J. Piskorz. An Overview of Fast Pyrolysis. In A. V. Bridgwater, editor, *Progress in thermochemical biomass conversion*, volume 2, pages 977–997. Blackwell Science, Oxford, 2001. ISBN 0-632-05533-2.

[186] T. Hosoya, H. Kawamoto, and S. Saka. Secondary reactions of lignin-derived primary tar components. *J. Anal. Appl. Pyrolysis*, 83(1):78–87, 2008.

[187] X. Bai, K. H. Kim, R. C. Brown, E. Dalluge, C. Hutchinson, Y. J. Lee, and D. Dalluge. Formation of phenolic oligomers during fast pyrolysis of lignin. *Fuel*, 128: 170–179, 2014.

[188] B. Scholze and D. Meier. Characterization of the water-insoluble fraction from pyrolysis oil (pyrolytic lignin). Part I. PY–GC/MS, FTIR, and functional groups. *J. Anal. Appl. Pyrolysis*, 60(1):41–54, 2001.

[189] N. Jendoubi, F. Broust, J. M. Commandre, G. Mauviel, M. Sardin, and J. Lédé. Inorganics distribution in bio oils and char produced by biomass fast pyrolysis: The key role of aerosols. *J. Anal. Appl. Pyrolysis*, 92(1):59–67, 2011.

[190] Q. Zhang, J. Chang, T. J. Wang, and Y. Xu. Review of biomass pyrolysis oil properties and upgrading research. *Energy Conversion and Management*, 48(1): 87–92, 2007.

[191] D. C. Elliott, A. Oasmaa, D. Meier, F. Preto, and A. V. Bridgwater. Results of the IEA Round Robin on Viscosity and Aging of Fast Pyrolysis Bio-oils: Long-Term Tests and Repeatability. *Energy & Fuels*, 26(12):7362–7366, 2012.

[192] T. Chen, C. Wu, R. Liu, W. Fei, and S. Liu. Effect of hot vapor filtration on the characterization of bio-oil from rice husks with fast pyrolysis in a fluidized-bed reactor. *Bioresour. Technol.*, 102(10):6178–6185, 2011.

[193] K. Raveendran and A. Ganesh. Heating value of biomass and biomass pyrolysis products. *Fuel*, 75(15):1715–1720, 1996.

[194] A. V. Bridgwater. Biomass Fast Pyrolysis. *Thermal Science*, 8(2):21–49, 2004.

[195] B. Pecha and M. Garcia-Perez. Chapter 26 - Pyrolysis of Lignocellulosic Biomass: Oil, Char, and Gas. In A. Dahiya, editor, *Bioenergy*, pages 413–442. Academic Press, Boston, 2015. ISBN 978-0-12-407909-0.

[196] A. Effendi, H. Gerhauser, and A. V. Bridgwater. Production of renewable phenolic resins by thermochemical conversion of biomass: A review. *Renew. Sust. Energ. Rev.*, 12(8):2092–2116, 2008.

[197] B. Sukhbaatar, P. H. Steele, and M. G. Kim. Use of lignin separated from bio-oil in oriented strand board binder phenol-formaldehyde resins. *Bioresources*, 4(2): 789–804, 2009.

[198] C. Amen-Chen, B. Riedl, X.-M. Wang, and C. Roy. Softwood Bark Pyrolysis Oil-PF Resols. Part 3. Use of Propylene Carbonate as Resin Cure Accelerator. *Holzforschung*, 56(3):281–288, 2002.

[199] Christian Hanser. *Strukturelle Untersuchungen, Reaktionen und Anwendung von Flash-Pyrolyseölen aus Biomasse.* PhD thesis, Universität Hamburg, Hamburg, 2002.

[200] M. M. Hossain, I. M. Scott, B. D. McGarvey, K. Conn, L. Ferrante, F. Berruti, and C. Briens. Toxicity of lignin, cellulose and hemicellulose-pyrolyzed bio-oil combinations: Estimating pesticide resources. *J. Anal. Appl. Pyrolysis*, 99:211–216, 2013.

[201] R. V. Pindoria, A. Megaritis, R. C. Messenbock, D. R. Dugwell, and R. Kandiyoti. Comparison of the pyrolysis and gasification of biomass: effect of reacting gas atmosphere and pressure on Eucalyptus wood. *Fuel*, 77(11):1247–1251, 1998.

[202] J. J. Órfão, F. J. A. Antunes, and J. L. Figueiredo. Pyrolysis kinetics of lignocellulosic materials - three independent reactions model. *Fuel*, 78(3):349–358, 1999.

[203] S. Sahoo, M. Ö. Seydibeyoğlu, A. K. Mohanty, and M. Misra. Characterization of industrial lignins for their utilization in future value added applications. *Biomass and Bioenergy*, 35(10):4230–4237, 2011.

[204] E. Jakab, O. Faix, F. Till, and T. Székely. Thermogravimetry/mass spectrometry study of six lignins within the scope of an international round robin test. *J. Anal. Appl. Pyrolysis*, 35(2):167–179, 1995.

[205] H. Abdullah and H. W. Wu. Biochar as a Fuel: 1. Properties and Grindability of Biochars Produced from the Pyrolysis of Mallee Wood under Slow-Heating Conditions. *Energy & Fuels*, 23(8):4174–4181, 2009.

[206] E. Suuberg, I. Aarna, and I. Milosavljevic. The Char Residues from Pyrolysis of Biomass – some Physical Properties of Importance. In A. V. Bridgwater, editor, *Progress in thermochemical biomass conversion*, volume 2, pages 1246–1258. Blackwell Science, Oxford, 2001. ISBN 0-632-05533-2.

[207] C. E. Brewer, K. Schmidt-Rohr, J. A. Satrio, and R. C. Brown. Characterization of Biochar from Fast Pyrolysis and Gasification Systems. *Environ. Prog. Sustain. Energy*, 28(3):386–396, 2009.

[208] M. Gupta, J. Yang, and C. Roy. Density of softwood bark and softwood char: procedural calibration and measurement by water soaking and kerosene immersion method. *Fuel*, 81(10):1379–1384, 2002.

[209] M. van de Velden, J. Baeyens, A. Brems, B. Janssens, and R. Dewil. Fundamentals, kinetics and endothermicity of the biomass pyrolysis reaction. *Renewable Energy*, 35(1):232–242, 2010.

[210] N. Prakash and K. T. Kinetic Modeling in Biomass Pyrolysis – A Review. *Journal of Applied Sciences Research*, 4(12):1627–1636, 2008.

[211] C. Di Blasi. Modeling and simulation of combustion processes of charring and non-charring solid fuels. *Prog. Energy Combust. Sci.*, 19(1):71–104, 1993.

[212] C. Di Blasi. Modeling chemical and physical processes of wood and biomass pyrolysis. *Prog. Energy Combust. Sci.*, 34(1):47–90, 2008.

[213] A. G. Liden, F. Berruti, and D. S. Scott. A kinetic model for the production of liquids from the flash pyrolysis of biomass. *Chemical Engineering Communications*, 65:207–221, 1988.

[214] Y. Haseli, J. A. van Oijen, and L. P. H. de Goey. Modeling biomass particle pyrolysis with temperature-dependent heat of reactions. *J. Anal. Appl. Pyrolysis*, 90(2):140–154, 2011.

[215] J. E. White, W. J. Catallo, B. L. Legendre, J. E. White, W. J. Catallo, and B. L. Legendre. Biomass pyrolysis kinetics: A comparative critical review with relevant agricultural residue case studies. *J. Anal. Appl. Pyrolysis*, 91(1):1–33, 2011.

[216] M. Al-Haddad, E. Rendek, J.-P. Corriou, and G. Mauviel. Biomass Fast Pyrolysis: Experimental Analysis and Modeling Approach. *Energy & Fuels*, 24:4689–4692, 2010.

[217] C. Serbanescu. Kinetic analysis of cellulose pyrolysis: a short review. *Chemical Papers*, 68(7):847–860, 2014.

[218] Bradbury, Allan G. W., Y. Sakai, and F. Shafizadeh. A kinetic model for pyrolysis of cellulose. *J. Appl. Polym. Sci.*, 23(11):3271–3280, 1979.

[219] A. Broido, M. Evett, and C. C. Hodges. Yield of 1,6-anhydro-3,4-dideoxy-β-d-glycero-hex-3-enopyranos-2-ulose (levoglucosenone) on the acid-catalyzed pyrolysis of cellulose and 1,6-anhydro-β-d-glucopyranose (levoglucosan). *Carbohydrate Research*, 44(2):267–274, 1975.

[220] C. A. Koufopanos, G. Maschio, and A. Lucchesi. Kinetic modelling of the pyrolysis of biomass and biomass components. *Can. J. Chem. Eng.*, 67(1):75–84, 1989.

[221] R. S. Miller and J. Bellan. A generalized biomass pyrolysis model based on super-imposed cellulose, hemicellulose and lignin kinetics. *Combustion Sc. & Tech*, 126 (1-6):97–137, 1997.

[222] C. Di Blasi and G. Russo. Modeling of Transport Phenomena and Kinetics of Biomass Pyrolysis. In A. V. Bridgwater, editor, *Advances in thermochemical biomass conversion*, volume 2, pages 906–921. Blackie Academic & Professional, London, 1994. ISBN 0751401730.

[223] M. L. Boroson, J. B. Howard, J. P. Longwell, and W. A. Peters. Product yields and kinetics from the vapor phase cracking of wood pyrolysis tars. *AIChE Journal*, 35 (1):120–128, 1989.

[224] B. Peters. Prediction of pyrolysis of pistachio shells based on its components hemicellulose, cellulose and lignin. *Fuel Process. Technol.*, 92(10):1993–1998, 2011.

[225] C. Couhert, J.-M. Commandre, and S. Salvador. Is it possible to predict gas yields of any biomass after rapid pyrolysis at high temperature from its composition in cellulose, hemicellulose and lignin? *Fuel*, 88(3):408–417, 2009.

[226] K. Raveendran, A. Ganesh, and K. C. Khilar. Pyrolysis characteristics of biomass and biomass components. *Fuel*, 75(8):987–998, 1996.

[227] F. P. Petrocelli and M. T. Klein. Simulation of Lignin Pyrolysis. *Chemical Engineering Communications*, 30(6):343–360, 1984.

[228] F. P. Petrocelli and T. M. Klein. Simulation of Kraft Lignin Pyrolysis. In R. Overend, T. Milne, and L. Mudge, editors, *Fundamentals of thermochemical biomass conversion*, pages 257–274. Elsevier Applied Science Publishers, London, New York, 1985. ISBN 0-85334-306-3.

[229] C. Dupont, L. Chen, J. Cances, J.-M. Commandre, A. Cuoci, S. Pierucci, and E. Ranzi. Biomass pyrolysis: Kinetic modelling and experimental validation under high temperature and flash heating rate conditions: Pyrolysis 2008 - Papers presented at the 18th International Symposium on Analytical and Applied Pyrolysis. *J. Anal. Appl. Pyrolysis*, 85(1-2):260–267, 2009.

[230] R. S. Miller and J. Bellan. Tar Yield and Collection from the Pyrolysis of Large Biomass Particles. *Combustion Sc. & Tech*, 127(1):97–118, 1997.

[231] D. S. Scott, J. Piskorz, M. A. Bergougnou, R. G. Graham, and R. P. Overend. The role of temperature in the fast pyrolysis of cellulose and wood. *Ind. Eng. Chem. Res*, 27(1):8–15, 1988.

[232] G. M. Simmons and M. Gentry. Particle size limitations due to heat transfer in determining pyrolysis kinetics of biomass. *J. Anal. Appl. Pyrolysis*, 10(2):117–127, 1986.

[233] C. Di Blasi. Kinetic and heat transfer control in the slow and flash pyrolysis of solids. *Ind. Eng. Chem. Res*, 35(1):37–46, 1996.

[234] D. Lathouwers and J. Bellan. Modeling and simulation of bubbling fluidized beds containing particle mixtures. *Proceedings of the Combustion Institute*, 28(2):2297–2304, 2000.

[235] D. Lathouwers and J. Bellan. Yield optimization and scaling of fluidized beds for tar production from biomass. *Energy & Fuels*, 15(5):1247–1262, 2001.

[236] Q. Xiong, S.-C. Kong, and A. Passalacqua. Development of a generalized numerical framework for simulating biomass fast pyrolysis in fluidized-bed reactors. *Chem. Eng. Sci.*, 99:305–313, 2013.

[237] Q. Xue, T. J. Heindel, and R. O. Fox. A CFD model for biomass fast pyrolysis in fluidized-bed reactors. *Chem. Eng. Sci.*, 66(11):2440–2452, 2011.

[238] Q. Xue, D. Dalluge, T. J. Heindel, R. O. Fox, and R. C. Brown. Experimental validation and CFD modeling study of biomass fast pyrolysis in fluidized-bed reactors. *Fuel*, 97:757–769, 2012.

[239] A. Trendewicz, R. L. Braun, A. Dutta, and J. Ziegler. One dimensional steady-state circulating fluidized-bed reactor model for biomass fast pyrolysis. *Fuel*, 133:253–262, 2014.

[240] N. Schwaiger, R. Feiner, H. Pucher, L. Ellmaier, J. Ritzberger, K. Treusch, P. Pucher, and M. Siebenhofer. BiomassPyrolysisRefinery – Herstellung von nachhaltigen Treibstoffen: BiomassPyrolysisRefinery – Production of Biofuels from Lignocellulose. *Chem. Ing. Tech*, 87(6):803–809, 2015.

[241] M. M. Wright, D. E. Daugaard, J. A. Satrio, and R. C. Brown. Techno-economic analysis of biomass fast pyrolysis to transportation fuels. *Techno-economic Comparison of Biomass-to-Biofuels Pathways*, 89, Supplement 1:S2–S10, 2010.

[242] Y. Zhang, T. R. Brown, G. Hu, and R. C. Brown. Techno-economic analysis of monosaccharide production via fast pyrolysis of lignocellulose. *Bioresour. Technol.*, 127(0):358–365, 2013.

[243] I. Petersen. *Sewage Sludge Gasification in the Circulating Fluidized Bed - Experiments and Modeling*. PhD thesis, Technische Universität Hamburg-Harburg, Hamburg, 2004.

[244] I. Petersen and J. Werther. Experimental investigation and modeling of gasification of sewage sludge in the circulating fluidized bed. *Chem. Eng. Process.*, 44(7):717–736, 2005.

[245] BSTFA + TMCS: Product Specification, 1997. URL http://www.sigmaaldrich.com/content/dam/sigma-aldrich/docs/Aldrich/General_Information/bstfa_tmcs.pdf.

[246] International Organization for Standardization. Particle size analysis – Laser diffraction methods, 2009.

[247] International Organization for Standardization. Particle size analysis – Image analysis methods – Part 2: Dynamic image analysis methods, 2006.

[248] DIN 66165. Siebanalyse, 1987.

[249] S. Willför, A. Pranovich, T. Tamminen, J. Puls, C. Laine, A. Suurnäkki, B. Saake, K. Uotila, H. Simolin, J. Hemming, and B. Holmbom. Carbohydrate analysis of plant materials with uronic acid-containing polysaccharides–A comparison between different hydrolysis and subsequent chromatographic analytical techniques. *Ind. Crops Prod.*, 29(2-3):571–580, 2009.

[250] D. Lorenz, N. Erasmy, Y. Akil, and B. Saake. A new method for the quantification of monosaccharides, uronic acids and oligosaccharides in partially hydrolyzed xylans by HPAEC-UV/VIS. *Carbohydrate polymers*, 140:181–187, 2016.

[251] K. S. W. Sing, D. H. Everett, R. A. W. Haul, L. Moscou, R. A. Pierotti, J. Rouquerol, and Siemieniewska T. Reporting physisorption data for gas/solid systems: with special reference to the determination of surface area and porosity. *Pure and Applied Chemistry*, 57(4):603–619, 1985.

[252] S. Lowell. *Characterization of porous solids and powders: Surface area, pore size and density*, volume 16 of *Particle technology series*. Kluwer Acad. Publ, Dordrecht, 4 edition, 2004. ISBN 9781402023026.

[253] DIN 51718. Prüfung fester Brennstoffe – Bestimmung des Wassergehaltes und der Analysenfeuchtigkeit, 06/2002.

[254] DIN 51719. Prüfung fester Brennstoffe – Bestimmung des Aschegehaltes, 07/1997.

[255] S. A. Channiwala and P. P. Parikh. A unified correlation for estimating HHV of solid, liquid and gaseous fuels. *Fuel*, 81(8):1051–1063, 2002.

[256] DIN 51777-1. Prüfung von Mineralöl-Kohlenwasserstoffen und Lösemitteln – Bestimmung des Wassergehaltes nach Karl Fischer – Direktes Verfahren, 03/1983.

[257] DIN 66137-1. Bestimmung der Dichte fester Stoffe – Teil 1: Grundlagen, 11/2003.

[258] S. Kabelac. *VDI-Wärmeatlas: Berechnungsunterlagen für Druckverlust, Wärme-und Stoffübergang*, volume 10., bearb. und erw. Aufl. Springer, Berlin, 2006. ISBN 3-540-25503-6.

[259] DIN 51605. Prüfung fester Brennstoffe – Bestimmung der Schüttdichte, 06/2001.

[260] V. Mathot, M. Pyda, T. Pijpers, G. Vanden Poel, E. van de Kerkhof, S. van Herwaarden, F. van Herwaarden, and A. Leenaers. The Flash DSC 1, a power compensation twin-type, chip-based fast scanning calorimeter (FSC): First findings on polymers: Special Issue: Interplay between Nucleation, Crystallization, and the Glass Transition. *Thermochim. Acta*, 522(1–2):36–45, 2011.

[261] A. Mettler-Toledo AG. Flash DSC 1 – Flash Differential Scanning Calorimetry: for Research and Development, 2010. URL http://us.mt.com/us/en/home/supportive_content/product_documentation/product_brochures/Flash_DSC_1/jcr:content/download/file/file.res/51725315_Flash_DSC1_Brochure_e.pdf.

[262] P. Cebe, B. P. Partlow, D. L. Kaplan, A. Wurm, E. Zhuravlev, and C. Schick. Using flash DSC for determining the liquid state heat capacity of silk fibroin. *Thermochim. Acta*, 615:8–14, 2015.

[263] J. Dykyj, J. Svoboda, R. C. Wilhoit, M. Frenkel, and K. R. Hall. Vapor Pressure of Chemicals Subvolume B Vapor Pressure and Antoine Constants for Oxygen Containing Organic Compounds, 2000.

[264] R. Patt, O. Kordsachia, and R. Süttinger. Pulp. In *Ullmann's Encyclopedia of Industrial Chemistry*, pages 477–540. Wiley-VCH Verlag GmbH & Co. KGaA, Weinheim and Germany, 2000. ISBN 3527306730.

[265] P. Tomani. The LignoBoost process. *Cell. Chem. Technol.*, 44(1-3):53–58, 2010.

[266] F. Öhman, H. Theliander, P. Tomani, and Axegård P. *Method for separating lignin from black liquor*. Patent, WO2006031175, 2006.

[267] C. Kirsch, K. Wörmeyer, C. Zetzl, and I. Smirnova. Enzymatische Hydrolyse von Lignocellulose im Festbettreaktor. *Chem. Ing. Tech*, 83(6):867–873, 2011.

[268] C. Kirsch, C. Zetzl, and I. Smirnova. Development of an integrated thermal and enzymatic hydrolysis for lignocellulosic biomass in fixed-bed reactors. *Holzforschung*, 65(4), 2011.

[269] C. Zetzl, K. Gairola, C. Kirsch, L. Perez-Cantu, and I. Smirnova. Ein-Reaktor-Konzept zur Hochdruckfraktionierung lignocellulosehaltiger Biomasse: One-Reactor Design for the Fractionation of Lignocellulosic Biomass under High Pressure. *Chem. Ing. Tech*, 84(1-2):27–35, 2012.

[270] L. Perez-Cantu, A. Schreiber, F. Schütt, B. Saake, C. Kirsch, and I. Smirnova. Comparison of pretreatment methods for rye straw in the second generation biorefinery: Effect on cellulose, hemicellulose and lignin recovery. *Bioresour. Technol.*, 142:428–435, 2013.

[271] W. Reynolds, C. Kirsch, and I. Smirnova. Thermal-Enzymatic Hydrolysis of Wheat Straw in a Single High Pressure Fixed Bed. *Chem. Ing. Tech*, 87(10):1305–1312, 2015.

[272] W. Reynolds, H. Singer, S. Schug, and I. Smirnova. Hydrothermal flow-through treatment of wheat-straw: Detailed characterization of fixed-bed properties and axial dispersion. *Chem. Eng. J.*, 281:696–703, 2015.

[273] A. Mayowa Azeez. *Enhancement of Chemical Products in Bio-Crude-Oil from Lignocellulosic Residues – Effects of Biomass Type, Temperature, Pre-treatment and Catalysts*. PhD thesis, Universität Hamburg, Hamburg, June 2011.

[274] T. Hatakeyama and H. Hatakeyama. *Thermal properties of green polymers and biocomposites*. Kluwer Academic Publishers, Dordrecht and Boston, 2004. ISBN 978-1-4020-2354-5.

[275] W. Glasser. Classification of Lignin According to Chemical and Molecular Structure. In *Lignin: Historical, Biological, and Materials Perspectives*, volume 742 of *ACS Symposium Series*, pages 216–238. American Chemical Society, 1999. ISBN 0-8412-3611-9.

[276] E. Zhuravlev and C. Schick. Fast scanning power compensated differential scanning nano-calorimeter: 2. Heat capacity analysis. *Thermochim. Acta*, 505(1–2):14–21, 2010.

[277] Quarzwerke GmbH. Quarzsand Frechen F 32 bis F 36. URL http://www.quarzwerke.at/datenblaetter/Quarzsand_F32-F36.pdf.

[278] A. Puettmann, E.-U. Hartge, and J. Werther. Application of the flowsheet simulation concept to fluidized bed reactor modeling. Part I: Development of a fluidized bed reactor simulation module. *Chem. Eng. Process.*, 60:86–95, 2012.

[279] M. Kramp. *Chemical Looping Combustion in Interconnected Fluidized Bed Reactors*. PhD thesis, Technische Universität Hamburg-Harburg, Hamburg, 2014.

[280] J. Werther and J. Wein. Expansion Behavior of Gas Fluidized Beds in the Turbulent Regime. *AIChE symposium series*, 90(301):31–44, 1994.

[281] J. Werther and E.-U. Hartge. A population balance model of the particle inventory in a fluidized-bed reactor/regenerator system. *Powder Technol.*, 148(2-3):113–122, 2004.

[282] J. Werther and E.-U. Hartge. Modeling of Industrial Fluidized-Bed Reactors. *Ind. Eng. Chem. Res*, 43(18):5593–5604, 2004.

[283] H. Bi. Some Issues on Core-Annulus and Cluster Models of Circulating Fluidized Bed Reactors. *Can. J. Chem. Eng.*, 80(5):809–817, 2002.

[284] W. C. Yang. Bubbling Fluidized Beds. In W.-C. Yang, editor, *Handbook of fluidization and fluid-particle systems*, volume 91 of *Chemical industries*, pages 54–113. Dekker, New York, 2003. ISBN 0203912748.

[285] J. M. D. Merry. Penetration of vertical jets into fluidized beds. *AIChE Journal*, 21 (3):507–510, 1975.

[286] C. Y. Wen, N. R. Deole, and L. H. Chen. A study of jets in a three-dimensional gas fluidized bed. *Powder Technol.*, 31(2):175–184, 1982.

[287] O. Levenspiel and D. Kunii. *Fluidization engineering*. Butterworth-Heinemann, Boston, 1991. ISBN 0-409-90233-0.

[288] K. Hilligardt. *Zur Strömungsmechanik von Grobkornwirbelschichten*. PhD thesis, Technische Universität Hamburg-Harburg, Hamburg, 1986.

[289] M. Schößler. *Mathematische Modellierung der Kohleverbrennung in technischen Wirbelschichten*. PhD thesis, Technische Universität Hamburg-Harburg, Hamburg, 1993.

[290] S. P. Sit and J. Grace. Effect of Bubble Interaction on Interphase Mass Transfer in Gas Fluidized Beds. *Chem. Eng. Sci.*, 36(2):327–335, 1981.

[291] B. E. Poling, J. P. O'Connell, and J. M. Prausnitz. *The properties of gases and liquids*. McGraw-Hill international editions Chemical engineering series. McGraw-Hill, New York, 5 edition, 2001. ISBN 9780071189712.

[292] W. Sitzmann, J. Werther, W. Böck, and G. Emig. Modelling of Fluidized Beds - Determination of suitable Kinetics for Complex Reactions. *German Chemical Engineering*, 8(5):301–307, 1985.

[293] J.-H. Choi, I.-Y. Chang, D.-W. Shun, C.-K. Yi, J.-E. Son, and S.-D. Kim. Correlation on the Particle Entrainment Rate in Gas Fluidized Beds. *Ind. Eng. Chem. Res*, 38 (6):2491–2496, 1999.

[294] W.-C. Yang, editor. *Handbook of fluidization and fluid-particle systems*, volume 91 of *Chemical industries*. Dekker, New York, 2003. ISBN 0203912748.

[295] I.-S. Han and C.-B. Chung. Dynamic modeling and simulation of a fluidized catalytic cracking process. Part I: Process modeling. *Chem. Eng. Sci.*, 56(5):1951–1971, 2001.

[296] H. Uhlemann and L. Mörl. *Wirbelschicht-Sprühgranulation*. Springer Berlin Heidelberg, Berlin, Heidelberg, 2000. ISBN 978-3-540-66985-2.

[297] W. Sitzmann, J. Werther, and G. Emig. Modellierung der Ethanol-Dehydratisierung im Wirbelschichtreaktor. *Chem. Ing. Tech*, 58(11):904–905, 1986.

[298] I. P. Boukis, M. Gyftopoulou, and I. Papamichael. Biomass Fast Pyrolysis in an Airblown Circulating Fluidized Bed Reactor. In A. V. Bridgwater, editor, *Progress in thermochemical biomass conversion*, volume 2, pages 1259–1267. Blackwell Science, Oxford, 2001. ISBN 0-632-05533-2.

[299] Chemical Composition of Wood, 2014. URL http://www.ipst.gatech.edu/faculty/ragauskas_art/technical_reviews/Chemical%20Overview%20of%20Wood.pdf.

[300] F. W. Hembach. *Verfahren zur Ermittlung charakteristischer Verbrennungsparameter einer Kohle in einer Labor-Wirbelschichtfeuerung.* PhD thesis, Technische Universität Hamburg-Harburg, Düsseldorf and VDI-Verlag, 1991.

[301] E. Biagini, P. Narducci, and L. Tognotti. Size and structural characterization of lignin-cellulosic fuels after the rapid devolatilization. *Fuel*, 87(2):177–186, 2008.

[302] K.-D. Henning and H. von Kienle. Carbon, 5. Activated Carbon. In *Ullmann's Encyclopedia of Industrial Chemistry*. Wiley-VCH Verlag GmbH & Co. KGaA, Weinheim and Germany, 2000. ISBN 3527306730.

[303] P. A. D. Rocca, E. G. Cerrella, P. R. Bonelli, and A. L. Cukierman. Pyrolysis of hardwoods residues: on kinetics and chars characterization. *Biomass and Bioenergy*, 16(1):79–88, 1999.

[304] N. Cao, H. Darmstadt, F. Soutric, and C. Roy. Thermogravimetric study on the steam activation of charcoals obtained by vacuum and atmospheric pyrolysis of softwood bark residues. *Carbon*, 40(4):471–479, 2002.

[305] M. Guerrero. Pyrolysis of eucalyptus at different heating rates: studies of char characterization and oxidative reactivity. *J. Anal. Appl. Pyrolysis*, 74(1):307–314, 2005.

[306] A. Aworn, P. Thiravetyan, and W. Nakbanpote. Preparation and characteristics of agricultural waste activated carbon by physical activation having micro- and mesopores. *J. Anal. Appl. Pyrolysis*, 82(2):279–285, 2008.

[307] M. Pointner, P. Kuttner, T. Obrlik, A. Jäger, and H. Kahr. Composition of corncobs as a substrate for fermentation of biofuels. *Agronomy Research*, 12(2):391–396, 2014.

[308] L. Wang, M. Yang, X. G. Fan, X. T. Zhu, T. Xu, and Q. P. Yuan. An environmentally friendly and efficient method for xylitol bioconversion with high-temperature-steaming corncob hydrolysate by adapted Candida tropicalis. *Process Biochemistry*, 46(8):1619–1626, 2011.

[309] C. E. Greenhalf, D. J. Nowakowski, A. B. Harms, J. O. Titiloye, and A. V. Bridgwater. A comparative study of straw, perennial grasses and hardwoods in terms of fast pyrolysis products. *Fuel*, 108:216–230, 2013.

[310] N. Jand and P. U. Foscolo. Decomposition of wood particles in fluidized beds. *Ind. Eng. Chem. Res*, 44(14):5079–5089, 2005.

[311] A. Tsutsumi, H. Suzuki, Y. Saito, K. Yoshida, and R. Yamazaki. Multi-component granulation in a fast fluidized bed. *Powder Technol.*, 100(2):237–241, 1998.

[312] H. Kage, M. Dohzaki, H. Ogura, and Y. Matsuno. Powder coating efficiency of small particles and their agglomeration in circulating fluidized bed. *Korean J. Chem. Eng.*, 16(5):630–634, 1999.

[313] H. Kage, R. Abe, R. Hattanda, T. Zhou, H. Ogura, and Y. Matsuno. Effect of solid circulation rate on coating efficiency and agglomeration in circulating fluidized bed type coater. *Powder Technol.*, 130(1-3):203–210, 2003.

[314] H. Kage, H. Nagasaki, A. Nitta, and M. Yamamura. Composition Effects of atomized suspension on efficiency, agglomeration and solid circulation rate of circulating fluidized bed coater. In Science and Technology, editor, *Fluidization*, pages 108–115, 2003.

[315] G. Xu, B. Stiller, E.-U. Hartge, and J. Werther. Prospect Demonstration of Spray Granulation in the Fast Fluidized Bed. In K. Cen, editor, *Proc. 8th Int. Conf. Circulating Fluidized Beds*, pages 809–818, 2005.

[316] B. Stiller, E.-U. Hartge, and J. Werther. Spray Granulation in the Circulating Fluidized Bed: Experimental Studies of Granulation Mechanisms in the Dense and Lean Phase. In J. Werther, W. Nowak, K.-E. Wirth, and E.-U. Hartge, editors, *Circulating Fluidized Bed Technology IX*, pages 471–476. TuTech Innovation GmbH, 2008. ISBN 978-3-930400-57-7.

[317] B. Stiller, E.-U. Hartge, S. Heinrich, and J. Werther. Spray Granulation in a Circulating Fluidized Bed System: The Role of Breakage and Attrition. In J. Li, F. Wei, X. Bao, and W. Wang, editors, *Proc. 11th Int. Conf. on Fluidized Bed Technol.*, pages 227–232. Chemical Industry Press, 2014. ISBN 978-7-122-20169-0.

[318] B. Stiller and J. Werther. *Sprühgranulation in der zirkulierenden Wirbelschicht*. PhD thesis, Technische Universität Hamburg-Harburg, Hamburg, 2016.

[319] J. Zank. *Tropfenabscheidung und Granulatwachstum bei der Wirbelschicht-Sprühgranulation*. PhD thesis, Universität Fridericiana Karlsruhe, Karlsruhe, 2003.

[320] B. Hirschberg and J. Werther. Factors affecting solids segregation in circulating fluidized-bed riser. *AIChE Journal*, 44(1):25–34, 1998.

[321] F. E. Brauns. *The Chemistry of Lignin*. Academic Press, 1952.

[322] M. Franck, E.-U. Hartge, S. Heinrich, J. Werther, P. Eidam, I. Fortmann, and D. Meier. Gewinnung von Phenolen aus Lignin durch Flash-Pyrolyse in einer zirkulierenden Wirbelschicht (ZWS) - Prozessentwicklung und Produktanalytik. In R. Abraham, editor, *Beiträge zur DGMK-Fachbereichstagung – Konversion von Biomassen*, volume 1, pages 35–44. Dt. Wiss. Ges. für Erdöl, Erdgas und Kohle (DGMK e.V.), 2012. ISBN 978-3-941721-24-1.

[323] A. Demirbas. The influence of temperature on the yields of compounds existing in bio-oils obtained from biomass samples via pyrolysis. *Fuel Process. Technol.*, 88(6):591–597, 2007.

[324] D. W. van Krevelen. Graphical-statistical method for the study of structure and reaction processes of coal. *Fuel*, 29:269–284, 1950.

[325] F. Brandt. *Brennstoffe und Verbrennungsrechnung*. Vulkan-Verlag GmbH, Essen, 3. edition, 1999. ISBN 9783802758010.

[326] J.G. Speight. *The Chemistry and Technology of Petroleum*. CRC Press, Boca Raton FL, 4 edition, 2007. ISBN 0-8493-9067-2.

[327] M. Asmadi, H. Kawamoto, and S. Saka. Thermal reactions of guaiacol and syringol as lignin model aromatic nuclei. *J. Anal. Appl. Pyrolysis*, 92(1):88–98, 2011.

[328] M. R. Rover, P. H. Hall, P. A. Johnston, R. G. Smith, and R. C. Brown. Stabilization of bio-oils using low temperature, low pressure hydrogenation. *Fuel*, 153:224–230, 2015.

[329] M. Berchtold, J. Fimberger, A. Reichhold, and P. Pucher. Upgrading of heat carrier oil derived from liquid-phase pyrolysis via fluid catalytic cracking. *Fuel Process. Technol.*, 142:92–99, 2016.

[330] P. Bielansky, A. Reichhold, and A. Weinert. Production of Gasoline and Gaseous Olefins: Catalytic Co-Cracking of Pyrolysis Oil Residue. In T. M. Knowlton, editor, *Proceedings of the Tenth International Conference on Circulating Fluidized Beds and Fluidization Technology - CFB-10*. Engineering Conferences International, 2011. ISBN 978-1-4507-7082-5.

[331] C. A. Mullen and A. A. Boateng. Catalytic pyrolysis-GC/MS of lignin from several sources. *Fuel Process. Technol.*, 91(11):1446–1458, 2010.

[332] T. N. Pham, D. Shi, and D. E. Resasco. Evaluating strategies for catalytic upgrading of pyrolysis oil in liquid phase. *Advances in Catalysis for Biomass Valorization*, 145: 10–23, 2014.

[333] E. Hoekstra, K. J. A. Hogendoorn, X. Wang, R. J. M. Westerhof, S. R. A. Kersten, W. P. M. van Swaaij, and M. J. Groeneveld. Fast Pyrolysis of Biomass in a Fluidized Bed Reactor: In Situ Filtering of the Vapors. *Ind. Eng. Chem. Res*, 48(10):4744–4756, 2009.

[334] A. V. Bridgwater. Upgrading biomass fast pyrolysis liquids. *Environ. Prog. Sustain. Energy*, 31(2):261–268, 2012.

[335] S. Czernik and A. V. Bridgwater. Overview of Applications of Biomass Fast Pyrolysis Oil. *Energy & Fuels*, 18(2):590–598, 2004.

[336] L. Zhang, R. H. Liu, R. Z. Yin, and Y. F. Mei. Upgrading of bio-oil from biomass fast pyrolysis in China: A review. *Renew. Sust. Energ. Rev.*, 24:66–72, 2013.

[337] P. M. Mortensen, J.-D. Grunwaldt, P. A. Jensen, K. G. Knudsen, and A. D. Jensen. A review of catalytic upgrading of bio-oil to engine fuels. *Appl. Catal. A Gen.*, 407 (1–2):1–19, 2011.

[338] Q. Bu, H. Lei, A. H. Zacher, L. Wang, S. Ren, J. Liang, Y. Wei, Y. Liu, J. Tang, Q. Zhang, and R. Ruan. A review of catalytic hydrodeoxygenation of lignin-derived phenols from biomass pyrolysis. *Bioresour. Technol.*, 124:470–477, 2012.

[339] Z. He and X. Q. Wang. Hydrodeoxygenation of model compounds and catalytic systems for pyrolysis bio-oils upgrading. *Catal. Sustain. Energy*, 1:28–52, 2012.

[340] X. Zhang, Q. Zhang, T. Wang, L. L. Ma, Y. Yu, and L. Chen. Hydrodeoxygenation of lignin-derived phenolic compounds to hydrocarbons over Ni/SiO2–ZrO2 catalysts. *Bioresour. Technol.*, 134:73–80, 2013.

[341] R. N. Olcese, J. Francois, M. M. Bettahar, D. Petitjean, and A. Dufour. Hydrodeoxygenation of Guaiacol, A Surrogate of Lignin Pyrolysis Vapors, Over Iron Based Catalysts: Kinetics and Modeling of the Lignin to Aromatics Integrated Process. *Energy & Fuels*, 27(2):975–984, 2013.

[342] A. M. Azeez, D. Meier, J. Odermatt, and T. Willner. Effects of zeolites on volatile products of beech wood using analytical pyrolysis. *J. Anal. Appl. Pyrolysis*, 91(2): 296–302, 2011.

[343] X. Meng, C. Xu, L. Li, and J. Gao. Kinetic Study of Catalytic Pyrolysis of C4 Hydrocarbons on a Modified ZSM-5 Zeolite Catalyst. *Energy & Fuels*, 24(12):6233–6238, 2010.

[344] D. J. Mihalcik, C. A. Mullen, and A. A. Boateng. Screening acidic zeolites for catalytic fast pyrolysis of biomass and its components. *J. Anal. Appl. Pyrolysis*, 92 (1):224–232, 2011.

[345] Z. Zhang, S. Sui, F. Wang, Q. Wang, and C. U. Pittman. Catalytic conversion of bio-oil to oxygen-containing fuels by acid-catalyzed reaction with olefins and alcohols over silica sulfuric acid. *Energies*, 6(9):4531, 2013.

[346] O. Bičáková and P. Straka. Production of hydrogen from renewable resources and its effectiveness. *Int. J. Hydrogen Energ.*, 37(16):11563–11578, 2012.

[347] H. Yang, R. Yan, H. Chen, D. H. Lee, and C. Zheng. Characteristics of hemicellulose, cellulose and lignin pyrolysis. *Fuel*, 86(12–13):1781–1788, 2007.

[348] K. V. Sarkanen and C. H. Ludwig, editors. *Lignins: Occurrence, formation, structure and reactions*. Wiley-Interscience, New York, 1971. ISBN 978-0471754220.

[349] J. Piskorz, D. Radlein, and D. S. Scott. On the mechanism of the rapid pyrolysis of cellulose. *J. Anal. Appl. Pyrolysis*, 9(2):121–137, 1986.

[350] Y. F. Liao, Wang, and X. Q. Ma. Study of reaction mechanisms in cellulose pyrolysis. *Abstr. Pap. Am. Chem. S.*, 227(Part 1):U1097, 2004.

[351] M. R. Hajaligol, J. B. Howard, J. P. Longwell, and W. A. Peters. Product compositions and kinetics for rapid pyrolysis of cellulose. *Ind. Eng. Chem. Proc. Des. Dev.*, 21(3):457–465, 1982.

[352] E. Butler, G. Devlin, D. Meier, and K. McDonnell. Characterisation of spruce, salix, miscanthus and wheat straw for pyrolysis applications. *Bioresour. Technol.*, 131:202–209, 2013.

[353] D. Chen, Y. Zheng, and X. Zhu. In-depth investigation on the pyrolysis kinetics of raw biomass. Part I: Kinetic analysis for the drying and devolatilization stages. *Bioresour. Technol.*, 131:40–46, 2013.

[354] A. Grimm, M. Öhman, T. Lindberg, A. Fredriksson, and D. Boström. Bed Agglomeration Characteristics in Fluidized-Bed Combustion of Biomass Fuels Using Olivine as Bed Material: Energy & Fuels. *Energy & Fuels*, 26(7):4550–4559, 2012.

[355] G. Varhegyi, B. Bobaly, E. Jakab, and H. Chen. Thermogravimetric Study of Biomass Pyrolysis Kinetics. A Distributed Activation Energy Model with Prediction Tests. *Energy & Fuels*, 25(1):24–32, 2011.

[356] M. Amutio, G. Lopez, R. Aguado, M. Artetxe, J. Bilbao, and M. Olazar. Kinetic study of lignocellulosic biomass oxidative pyrolysis. *Fuel*, 95:305–311, 2012.

[357] T. Chen, C. Deng, and R. Liu. Effect of Selective Condensation on the Characterization of Bio-oil from Pine Sawdust Fast Pyrolysis Using a Fluidized-Bed Reactor. *Energy & Fuels*, 24(12):6616–6623, 2010.

[358] J. Werther and E.-U. Hartge. Elutriation and Entrainment. In W.-C. Yang, editor, *Handbook of fluidization and fluid-particle systems*, volume 91 of *Chemical industries*, pages 113–128. Dekker, New York, 2003. ISBN 0203912748.

[359] L. Ratschow. *Three-dimensional simulation of temperature distributions in large-scale circulating fluidized bed combustors*. PhD thesis, Hamburg University of Technology, Aachen, 2009.

[360] R. Wischnewski. *Simulation of large-scale circulating fluidized bed combustors*. PhD thesis, Technische Universität Hamburg-Harburg, Shaker and Aachen, 2009.

[361] Arthur, J. R. Reactions between carbon and oxygen. *Trans. Faraday Soc.*, 47: 164–178, 1951.

[362] I. B. Ross and J. F. Davidson. The Combustion of Carbon Particles in a Fluidized Bed. *Trans. Instn Chem. Engrs*, 60(2):108–114, 1982.

[363] J. B. Howard, G. C. Williams, and D. H. Fine. Kinetics of carbon monoxide oxidation in postflame gases. *14th Symp. (Int.) on Comb.*, 14(1):975–986, 1973.

A APPENDIX

A.1 Figures

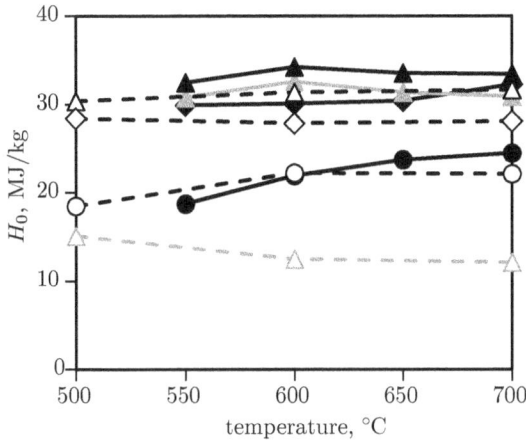

Figure A.1: HHV of pyrolysis products for Kraft lignin (filled symbols) and hyrolysis lignin (empty symbols), char (daf) ▲, char (dry) ▲, oil (daf) ◇, gas ●
char and oil H_0 calculated by Eq. 4.1 and gas H_0 determined by tabulated values and volumetric gas composition

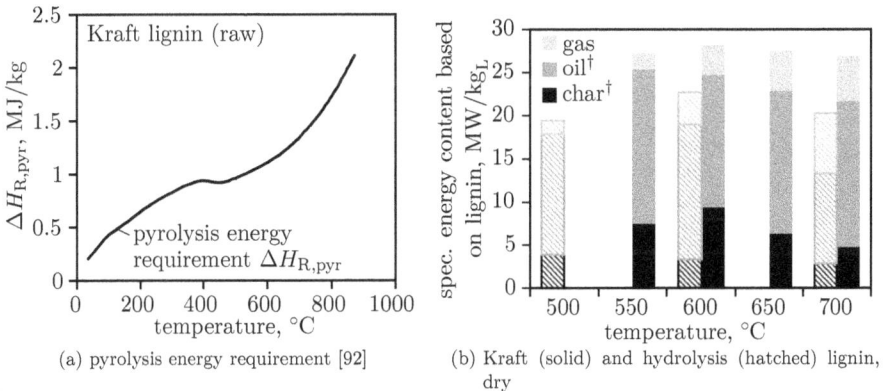

(a) pyrolysis energy requirement [92]

(b) Kraft (solid) and hydrolysis (hatched) lignin, dry

Figure A.2: Energy balance related values for CFB lignin pyrolysis
[†]calculated by $H_{0,i} \cdot Y_i$ and $H_{0,i}$ from Eq. 4.1

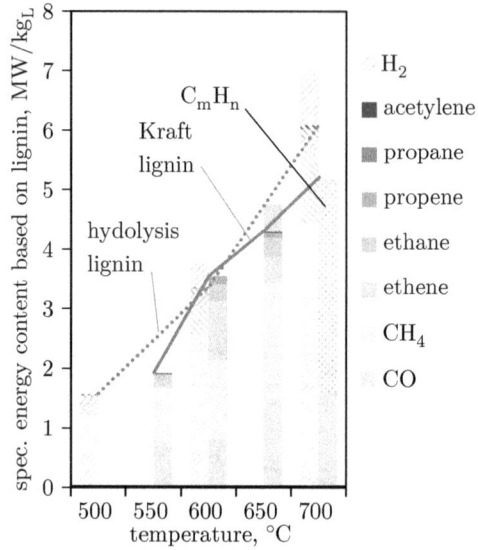

Figure A.3: Specific energy content of gas components based on dry lignin for Kraft (solid) and hydrolysis (hatched) lignin

calculated by $H_{0,i} \cdot Y_i$ with tabulated values for $H_{0,i}$

A.2 Tables

Table A.1: Input parameter for modeling of SPE circulating fluidized bed pyrolysis plant

parameter	value	unit / reference
gas distributor	technical	-
number of orifices per cross-sectional area	596	$1/m^2$
density active (solid density wood)	1530	kg/m^3
density char	1600	kg/m^3
density quartz sand	2300	kg/m^3
number of discretization elements	400	-
pressure	1.2	bar
biomass feed	2	kg/h
gas feed	86.1	m^3/h
solids inventory of bed	0.7	kg
solids folume fraction at minimal fluidization	0.4	m^3/m^3
Geldart class	B	-
reactor height	2	m
reactor diameter	80	mm
correlation for minimal fluidization velocity	Richardson	[287]
correlation for TDH	Chan & Knowlton	[358]
max. bubble diameter	80	mm
Sauter diameter quartz sand	200	µm

Table A.2: Input parameter for modeling of circulating fluidized bed and bubbling fluidized bed pyrolysis plants of CERTH and University of Waterloo, respectively

parameter	unit	CFB (CERTH)	BFB (WATERLOO)
pressure	bar	1.2	1.25
biomass feed[†]	kg/h	9.65	2.445[‡], 1.768[*], 1.765[*]
solids inventory of bed	kg	0.17	1
solids volume fraction of bed	-	0.4	0.5
reactor height	mm	2000	407
reactor diameter	mm	50	101
superficial gas velocity[†]	m/s	3.53	0.814[‡], 0.674[*], 0.729[*]
biomass composition	cf. Table	5.7	5.9
max. bubble diameter	mm	50	101
sauter diameter sand	µm	200	300

[†]average values for all experiments with the same biomass, for biomass: [‡]maple wood, [*]wheat straw and [*]aspen wood

Table A.3: List of samples and corresponding operating conditions

sample no.	lignin type	$\dot{V}_{N_2}$ (s.c.) m³/h	temperature			ϑ_{SS} min	lignin feed		overpressure			online concentration			
			$\vartheta^\dagger$ °C	ϑ_{19} °C	ϑ_{SS} °C		$m_{L,SS}$ kg	$\dot{m}_L$ kg/h	p_{19} mbar	p_{12} mbar	p_{SS} mbar	CO vol.-%	CO_2 vol.-%	CmHn vpm	O_2,ss vol.-%
57B	KL	17.0	647	53	28	17.5	0.457	1.55	4.3	49	10	1.03	0.33	7455	0.3
58A	KL	17.0	648	55	28	12.4	0.353	1.70	3.5	42	16	1.13	0.34	8104	1.6
58B	KL	17.0	588	57	29	9.5	0.243	1.52	1.1	41	13	0.97	0.36	8202	2.5
61A	KL	14.7	650	60	31	13.0	0.489	2.23	1.3	49	15	1.51	0.89	13906	0.3
61B	KL	14.7	654	64	31	9.5	0.341	2.25	1.2	41	16	1.43	0.99	13843	0.5
63A	KL	15.9	602	58	31	16.5	0.579	2.09	4.8	45	19	0.93	0.54	6837	0.0
63B	KL	15.9	601	57	31	16.5	0.606	2.22	5.0	41	18	0.94	0.62	7373	1.7
64A	KL	14.7	704	57	29	10.2	0.234	1.39	3.9	44	13	1.15	0.49	9178	0.1
64B	KL	14.7	701	57	30	4.4	0.131	1.46	4.1	45	16	1.20	0.49	9444	0.8
68A	KL	10.9	665	62	27	12.8	0.348	1.63	1.6	48	10	1.45	0.51	17031	0.0
68B	KL	10.9	650	59	28	6.0	0.206	2.06	2.0	43	12	1.83	0.64	21546	0.5
69A	KL	10.9	661	61	29	8.8	0.252	1.73	1.4	45	10	1.56	0.77	11421	0.0
69B	KL	10.9	659	63	30	7.6	0.230	1.81	1.5	45	13	1.99	0.90	15111	1.9
70A	KL	16.8	655	59	32	9.0	0.506	3.39	4.8	47	12	2.32	0.54	13881	0.0
70B	KL	16.8	655	58	32	8.7	0.565	3.89	4.9	49	13	2.87	0.71	17029	0.4
71A	KL	28.4	656	45	29	11.3	0.301	1.60	25.7	42	65	0.55	0.21	3523	0.0
71B	KL	28.4	658	45	29	7.5	0.211	1.70	24.8	47	34	0.67	0.23	4238	1.4
72A	KL	28.4	658	44	28	9.3	0.279	1.81	27.8	41	32	0.96	0.26	3921	0.0
72B	KL	28.4	651	44	29	9.4	0.286	1.83	24.9	50	20	0.98	0.30	4389	0.0
73A	KL	17.0	557	52	28	10.2	0.324	1.90	8.0	46	19	0.75	0.38	3634	0.0
73B	KL	17.0	559	52	29	7.0	0.252	2.16	7.6	46	38	0.78	0.42	4060	0.0
78A	KL	17.6	656	53	28	8.4	0.516	3.68	18.6	54	14	1.91	0.79	12727	0.4
78B	KL	17.6	658	54	28	8.1	0.265	1.95	18.3	60	13	1.05	0.50	7678	0.0
78C	KL	17.6	658	54	29	8.1	0.238	1.77	18.1	80	24	0.99	0.53	7628	0.0
78D	KL	17.6	665	55	29	11.8	0.366	1.87	17.9	134	16	1.27	0.65	9763	0.4

Table A.3: (continued)

sample no.	lignin type	$\dot{V}_{N_2}$ m³/h (s.c.)	temperature			lignin feed			overpressure			online concentration			
			$\vartheta^\dagger$ °C	ϑ_{19} °C	ϑ_{SS} °C	t_{SS} min	$m_{L,SS}$ kg	$\dot{m}_L$ kg/h	p_{19} mbar	p_{12} mbar	p_{SS} mbar	CO vol.-%	CO_2 vol.-%	CmHn vpm	$O_{2,SS}$ vol.-%
79A	KL	15.9	648	48	26	9.3	0.299	1.94	22.4	55	14	1.05	12.70	7029	0.0
79B	KL	15.9	656	50	27	8.1	0.274	2.02	22.6	60	12	1.14	12.86	7783	0.0
81A	KL	17.6	651	50	27	10.0	0.366	2.19	22.3	32	13	1.09	0.27	7042	0.1
81B	KL	17.6	656	52	28	10.6	0.362	2.04	22.6	32	12	1.07	0.31	7266	0.0
81C	KL	17.6	655	50	29	10.5	0.347	1.99	22.3	35	12	1.06	0.39	7657	0.0
81D	KL	17.6	657	53	29	9.8	0.296	1.79	21.6	54	13	1.04	0.40	7455	0.0
87A	HL	28.3	599	56	23	4.9	0.297	3.60	53.0	135	11	1.26	0.73	n.d.	0.0
87B	HL	28.3	597	56	24	9.0	0.552	3.70	52.8	133	8	0.85	0.45	4390	0.0
88A	HL	26.9	596	49	20	6.6	0.400	3.65	50.6	133	13	1.18	0.52	7664	1.0
88B	HL	26.9	596	56	20	13.0	0.810	3.73	48.5	140	10	1.27	0.60	8316	0.2
88C	HL	26.9	597	59	22	10.0	0.624	3.74	51.7	147	10	1.20	0.58	8651	0.0
89A	KL	26.9	651	60	24	14.7	0.509	2.08	51.4	138	9	0.72	0.19	5377	0.0
89B	KL	26.9	651	58	25	13.0	0.434	2.01	49.1	129	13	0.63	0.20	5291	0.9
89C	KL	26.8	648	59	25	13.7	0.537	2.36	48.3	152	10	0.71	0.26	6309	0.0
89D	KL	25.8	649	57	25	13.0	0.511	2.36	46.2	147	11	0.74	0.28	6472	0.0
90A	HL	24.6	694	55	24	12.8	0.886	4.14	39.5	121	12	3.20	0.62	15725	0.0
90B	HL	24.6	699	56	24	5.5	0.417	4.55	40.1	127	17	3.56	0.71	17553	0.7
90C	HL	21.7	692	61	25	8.0	0.618	4.64	32.0	179	10	3.98	0.89	20277	0.0
91A	HL	28.7	497	43	22	8.3	0.473	3.42	43.9	134	14	0.47	0.36	2623	0.1
91B	HL	28.7	500	44	22	14.7	0.942	3.85	46.3	141	10	0.58	0.43	3590	0.0
91C	HL	26.6	499	48	22	14.0	0.986	4.23	41.0	157	10	0.82	0.63	5311	0.0

$\dagger$mean riser temperature

A.3 Char combustion reactions

A.3.1 Heterogeneous char combustion

The biomass char combustion reaction is described by a simplified model which has proven high accuracy in coal char combustion [279, 359, 360]. For a detailed description it is referred to [279]. Devolatilization, as e.g. described by Kramp [279], is neglected due to the assumption of sufficient devolatilization during pyrolysis. The oxidation of char to CO and CO_2 can be combined in one equation as

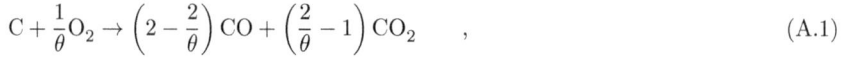

$$C + \frac{1}{\theta}O_2 \rightarrow \left(2 - \frac{2}{\theta}\right)CO + \left(\frac{2}{\theta} - 1\right)CO_2 \quad , \tag{A.1}$$

wherein θ is mechanism factor, which determines the ratio between the main combustion products CO and CO_2.

$$\theta = \begin{cases} \frac{2p^*+2}{p^*+2} & d_p < 50\,\mu m \\ \frac{2p^*+2-p^*\left(\frac{d_p-50\,\mu m}{950\,\mu m}\right)}{p^*+2} & 50\,\mu m \leq d_p < 1000\,\mu m \\ 1 & d_p \leq 1000\,\mu m \end{cases} \tag{A.2}$$

p^* is the ratio of produced CO to produced CO_2 and can be determined by an empirical correlation [289, 361]:

$$p^* = 2500 \cdot \exp\left[-\frac{6240K}{T_p}\right] \quad . \tag{A.3}$$

Due to the released combustion heat, the single char particle temperature T_p is higher than the bed temperature T_{bed} [362]:

$$T_p = T_{bed} + 6.6\left[m^3 \cdot K/mol\right] \cdot c_{O_2,m}\left[mol/m^3\right] \tag{A.4}$$

A.3.2 Homogeneous gas phase reaction

The reaction rate constant for the homogeneous reaction of

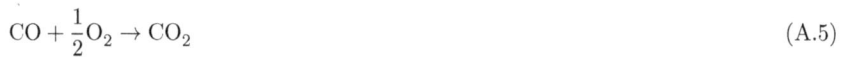

$$CO + \frac{1}{2}O_2 \rightarrow CO_2 \tag{A.5}$$

was determined by Howard et al. [363]

$$k_{CO} = 1.3 \cdot 10^{11} \exp\left[\frac{-15088\,K}{T_{bed}}\right] \tag{A.6}$$

and the homogeneous reaction rate [363] is:

$$r_{CO} = k_{CO} \cdot c_{CO}\sqrt{c_{O_2} \cdot c_{H_2O}} \quad . \tag{A.7}$$

Further details can also be read in [300].

A.4 Student projects

Table A.4: Student projects under the authors' supervision that contributed to this work

student	thesis title	year	thesis type
P. Clauß	Modellierung der Ligninpyrolyse in einer Wirbelschicht	2011	master
H. Evers	Lignin-Pyrolyse: Experimentelle Bestimmung der Produktausbeute	2011	bachelor
T. Voß	Ligninpyrolyse – Einfluss von Betriebsbedingungen aus die Produktcharakteristik	2011	project
T. Becke	Experimentelle Ligninpyrolyse: Untersuchung der Produktzusammensetzung bei verschiedenen Prozessbedingungen	2012	bachelor
L. Groos	Experimentelle Ausbeuteuntersuchung der Ligninpyrolyse	2012	bachelor
A.-L. Bologna	Pyrolyse von Lignin in einem Laborreaktor	2013	bachelor
D. Kant	Modellierung einer zirkulierenden Wirbelschicht im Aspen Custom Modeler	2013	project
A. Krenz	Konzeption und Inbetriebnahme eines Laborreaktors für die Pyrolyse von Lignin zur Analyse der Produktzusammensetzung	2013	bachelor
S. Sauerschell	Experimentelle Untersuchung der Lignin-Pyrolyse: Abhängigkeit der Produktausbeute von den Betriebsbedingungen	2013	master
S. Chernikov	Experimentelle Bestimmung von Massen- und Elementbilanzen bei der Pyrolyse von Lignin	2014	bachelor
M. Holz	Modellierung der Biomassepyrolyse in einer zirkulierenden Wirbelschicht	2014	master
M. Hug	Modellierung der Verbrennung von Feststoff in einer stationären Wirbelschicht	2014	bachelor
J.-M. Bettien	Pyrolyse von Kraft-Lignin in der zirkulierenden Wirbelschicht – Vergleich zweier Anlagenkonfigurationen	2015	bachelor
N. Ellenfeld	Modellierung einer Wirbelschichtverbrennung – Betrachtung von Hydrodynamik, Stoff- und Energiebilanz	2015	project
J. Faass	Thermische Analyse von Biomassekomponenten im Hinblick auf die Pyrolyse in der zirkulierenden Wirbelschicht	2015	bachelor
J.-P. Kreienborg	Modellierung der Biomassepyrolyse in einem Gaszyklon	2015	bachelor
T. Wytrwat	Inbetriebnahme einer Anlage zur experimentellen Untersuchung der Ligninpyrolyse	2015	master
K. Gescher	Modellierung der Pyrolyse von Biomasse mit integriertem Koksabbrand in gekoppelten Wirbelschichten	2016	master

Curriculum Vitae

Miika Franck

March 11th, 1983	born in Hamburg, Germany

Experience

since 05/2016	Research Engineer at BASF SE, Ludwigshafen, Germany
09/2010 – 11/2015	Research Fellow at the Institute of Solids Process Engineering and Particle Technology, Hamburg University of Technology, Germany

Academic History

10/2003 – 08/2010	Studies of Energy and Environmental Engineering at Hamburg University of Technology, Germany Qualification: Diplom-Ingenieur
10/2007 – 03/2008	Studies at Tongji University, Shanghai, China

Education

08/1994 – 06/2003	Alexander-von-Humboldt-Gymnasium, Hamburg, Germany Qualification: Abitur
07/2000 – 06/2001	Quincy Senior High School, Quincy IL, USA
09/1990 – 07/1994	Elementary School, Grundschule Kanzlerstraße, Hamburg, Germany